T0146104

delta of power

Technology in Motion

Pamela O. Long and Asif Siddiqi, Series Editors

Published in cooperation with the Society for the History of Technology (SHOT), the Technology in Motion series highlights the latest scholarship on all aspects of the mutually constitutive relationship between technology and society. Books focus on discrete thematic or geographic areas, covering all periods of history from antiquity to the present around the globe. These books synthesize recent scholarship on urgent topics in the history of technology with a sensitivity to challenging perspectives and cutting-edge analytical approaches. In combining historical and historiographical approaches, the books serve both as scholarly works and as ideal entry points for teaching at multiple levels.

delta of power

the military-industrial complex

Alex Roland

Johns Hopkins University Press
Baltimore

© 2021 Johns Hopkins University Press
All rights reserved. Published 2021
Printed in the United States of America on acid-free paper
9 8 7 6 5 4 3 2 1

Johns Hopkins University Press
2715 North Charles Street
Baltimore, Maryland 21218-4363
www.press.jhu.edu

Library of Congress Cataloging-in-Publication Data

Names: Roland, Alex, 1944– author.
Title: Delta of power : the military-industrial complex / Alex Roland.
Description: Baltimore : Johns Hopkins University Press, 2021. |
   Series: Technology in motion | Includes bibliographical references and index.
Identifiers: LCCN 2020047878 | ISBN 9781421441818 (paperback) |
   ISBN 9781421441825 (ebook)
Subjects: LCSH: Military-industrial complex—United States.
Classification: LCC HC110.D4 R65 2021 | DDC 338.4/735500973—dc23
LC record available at https://lccn.loc.gov/2020047878

A catalog record for this book is available from the British Library.

The index was written by Victoria George, PhD.

Special discounts are available for bulk purchases of this book. For more
information, please contact Special Sales at specialsales@jh.edu.

Johns Hopkins University Press uses environmentally friendly book materials,
including recycled text paper that is composed of at least 30 percent post-
consumer waste, whenever possible.

Contents

Charts and Figures

Acknowledgments

Of the many friends and colleagues who helped make this book possible, none is more deserving of recognition than Pamela Long. She co-edited the series Historical Perspectives on Technology, Society, and Culture in which the first version of this book appeared, a joint venture of the American Historical Association and the Society for the History of Technology. And she helped to negotiate the agreement by which that series was transferred to Johns Hopkins University Press and reborn as the Technology in Motion series. She partnered with co-editors Robert C. Post in the first series and Asif Siddiqi in the current effort, both friends and colleagues with whom I have been similarly honored to work. In all three cases, their contributions ranged from recruitment through conceptualization to reading and commenting.

Four colleagues read this complete manuscript in draft and made invaluable comments and suggestions: T. S. Allen, Arne Axelsson, Wayne E. Lee, and Mark Wilson. Their advice and encouragement made this a better book than it would have been.

The Duke University Libraries, especially the William R. Perkins and Roy J. Bostock Libraries, supported this project in ways too numerous to recount. In both resources and staff, these bountiful libraries found, acquired, copied, or borrowed on my behalf published and unpublished sources ranging from archival and government documents to electronic records and data. This project, and the pandemic that complicated it, reminded me of what a powerful tool the internet is, especially for a lone scholar researching from isolation at home. But a magnificent library and an indefatigable staff can still work miracles. Too many people at the Duke Libraries did too many services, some anonymously, to recognize them all. But I am particularly grateful to data visualization analyst Eric Monson for converting tabular data into the charts that grace this text. Some of these, I think, made my points far more effectively than my words.

The staff at Johns Hopkins University Press were pleasant and helpful from start to finish of this project. Victoria George, like me a Duke retiree, synthesized an index that serves as a true portal to the book and its concepts.

Finally, my wife, Liz, as always, suffered through months of my isolation, preoccupation, and mood swings. As always, she offered just the right combination of encouragement and home truths to keep me on course and on time. She makes my books better and my life richer.

Even with all this help, however, this book doubtless retains mistakes, for which I am solely responsible.

delta of power

Introduction

This book posits that the Military-Industrial Complex of the Cold War was unique.[1] It addresses a single question: did that Military-Industrial Complex (MIC) continue into the subsequent three decades? Does the United States still have an MIC or has the United States' relationship with its arms manufacturers changed beyond recognition? If the MIC is no longer recognizable, what has replaced it?

This volume expands upon an earlier study of the MIC, published on the eve of the terrorist attacks on the United States on September 11, 2001.[2] That transformative event in United States and world history jolted Americans from their brief hegemony in a unipolar world. Suddenly, the United States found itself committed to an unfamiliar, asymmetric contest with global terrorism while trying to resist the emergence of peer rivals. Terrorism required new strategies, institutions, and paths to innovation. It was not immediately clear if the old MIC, which had won the Cold War, would prove adequate to these new challenges. Perhaps it is still not clear. But these are the questions that this book attempts to answer.

This task will require a clearer definition of the Military-

Industrial Complex than was provided in the 2001 edition. Although most works that invoke any military-industrial complex neglect to provide a definition, some students of the subject have attempted to specify what they mean by the term. J. Paul Dunne and Elisabeth Sköns, for example, note that during the Cold War, the MIC was generally taken to mean "the vested interests within the state and industry in expanding the military sector and in increasing military spending, with external threats providing the justification."[3] Economist Ron Smith sees the MIC as "coalitions of vested interests ... [that] shape choices against peaceful goals and national security interests in order to extract funds for their own purposes." He adds that "the national arms industry can become a political actor in its own right."[4] Drawing on the insights of these and other scholars, this work defines the US Military-Industrial Complex of the Cold War as *a convergence of state and industrial forces collaborating to shape US national security policy to privilege the forces' respective special interests over the national interest and national security.*

President Dwight D. Eisenhower had a very specific phenomenon in mind when he introduced the term in his farewell address on January 17, 1961. Perhaps the most consistent and abiding feature of the term "Military-Industrial Complex" is the pejorative flavor that Eisenhower imparted to it.[5] Though the term has been used to label phenomena as benign as America's arsenal of democracy in World War II, it nonetheless conveys opprobrium, suspicion, and danger.[6] Indeed, Eisenhower's farewell address is often portrayed as a "warning," delivered from the depths of the Cold War, exhorting the American people to beware of perils within their own ranks. It is necessary, therefore, to define the term broadly and to appreciate the many dimensions of the Cold War MIC that Eisenhower and his successors have experienced. These variables will also aid in comparing Eisenhower's MIC with other relationships between the American government and its defense industry in the thirty years since the end of the Cold War.

This study is organized in two parts. The first part reprises the analysis in the 2001 version of this book, but with minor changes to that text and the updating and expansion of the references. Part II of this study follows with a very different analysis. It is chronological and narrative, while part I is more topical and conceptual. There is an epistemological reason for this difference in style and format. In historical retrospect, the period of the Cold War seems all of a piece. Historical change certainly took place during the Cold War, but that period had a military, economic, and political unity that invited generalization. In contrast, the three decades since the end of the Cold War in 1991 have witnessed less cohesion, faster change, and more historical contingency. Even after thirty years, it is difficult to define America's military-industrial institutions and infrastructure, let alone suggest where they may be trending. But that more contingent and evolving story nonetheless helps to reveal the Cold War MIC in better perspective and to compare it to what followed. Those comparisons will be highlighted in the conclusion.

Despite the different narrative structures of parts I and II, this entire study coheres around five institutional variables—the state, strategy, the military, industry, and contracts—and four themes—technological determinism, economic impact, statism, and collusion.

First and foremost among the institutional variables is *the state*. The United States is and has always been a constitutional, representative democracy, hosting free-enterprise capitalism. The United States has always imposed upon its elected representatives in Congress dual allegiances, to the voters who elected them and to the Constitution they are sworn to uphold. Throughout its history, the US has found itself pulled between centripetal forces of statism, gathering power in a large federal government, and centrifugal forces of state and local governments exercising the powers not otherwise granted to the national authority. The struggle has been Herculean and inconclusive. Ever since World War II, the US has been the strongest economic and military power in the world. Since its founding, it enjoyed free security from great-power threats provided by the Atlantic and Pacific Oceans, an advantage it lost to military technology in World War II. Since that historic conflict, the United States has pursued collective security with allies around the globe and formulated, institutionalized, and defended a liberal world order articulated in the United Nations Charter.

The second variable is *strategy*. Like most states, the United States has prioritized protection of its territory, people, and interests "from all enemies, foreign and domestic."[7] During the Cold War, Americans generally perceived the greatest threat to be the Soviet Union and a worldwide communist movement. Indeed, they saw this as an existential threat. They responded with three strategies. First, they committed the United States to "containment" of world communism. Second, they based their military strength on the quality—not the quantity—of their military forces, betting that superior technology could counterbalance the numerical superiority of Soviet armed forces. Third, they organized a series of alliances with like-minded states—mostly western democracies—in what they called the "free world." This strategy ignited a nuclear arms race, based on deterrence, paired with a succession of armed interventions with conventional weapons to blunt communist interference in other states.

The *military* is the third variable in the US MIC. From the founding of the nation, the United States has institutionalized and enforced civilian control of the military. Numerous provisions of the Constitution embody this principle, and long practice has embedded it deeply in American culture. Still, World War II inaugurated a sea change in America's relationship with its military. At the end of that war, unlike all before it, the United States began to maintain a large, standing military establishment in peacetime. Without this, there would have been no MIC. The size and duration of this establishment also gave rise to other developments that transformed the relationship between the US and its armed forces. Attempts by civilian leaders to limit the expense of these forces drove the services into competition with each other for limited funding. Each service sought roles and missions to expand its claims on resources. For a while, the air force commanded the lion's share, contributing to the rise of the aerospace realm as the dominant driver of the American arms industry.[8] And the size of each service's arsenal, in turn, drove the number of personnel required to maintain and operate their tools of war.[9] Additionally, other government agencies—the Atomic Energy Commission / Department of Energy, the Central Intelligence Agency, and the National Aeronautics and Space Administration, for example—took on functions supporting the national military establishment, increasing both the cost of security and the variety of actors seeking to contribute.

Just as the military failed to demobilize after World War II, so, too, did the vast arms *industry* assembled in that conflict. It founded a defense industrial base (DIB), an oligopoly tailored to satisfy military demand for the tools of war. Continuing another World War II innovation, the government elected to procure both research and development (R&D) and its instruments of war through contracts with the DIB, greatly reducing its pre-war dependence on its own arsenals and shipyards. Through most of the Cold War, however, the United States resisted national industrial policy, the practice of communist, socialist, and even some capitalist states to promote select industries with government subsidies and preference. The state did, however, allow some products of the American arms industry to be sold abroad under certain conditions. These sales supported the industry, recovered R&D investments, and lowered the unit cost of American purchases. Over time, the state sought to promote dual-use technologies with applications in both the military and private sectors and to monitor the "revolving door" through which individuals moved back and forth between government and industry positions.

Contracts, the instruments by which the government purchased goods and services from the private sector, in time became an institution in their own right. The government and its contractors shared a principal-client relationship of cooperation and contention. Both wanted the US armed forces to have the best arms and equipment in the world. But the state wanted to maximize savings and capabilities, and the industry wanted to maximize profits. The search for a contract that would allow the government to achieve its ends while sustaining an industrial base vexed generations of government procurement officers and provided perverse incentives for industry to cheat. The process was complicated by the uncertainties of contracting for innovation; the long lead times often involved in the development and manufacture of large-scale technological systems; the constraints of secrecy; and the peculiarities of a marketplace dominated by few sellers and still fewer buyers. The government experimented with a variety of contractual forms designed to specify the unknown, routinize innovation, and constrain human nature. All the while, the government sought to manage the DIB as a whole without slipping into national industrial policy. Sometimes the military rescued floundering companies to maintain competition and production backup; at other times, it allowed—even encouraged—industry consolidation in the interests of economy, resilience, and efficiency. At the same time, industry sought guarantees of survival in a volatile and unpredictable market. In short, the MIC *was* a national industrial policy, if an inadvertent one.

These five institutions—the state, strategy, the military, industry, and contracts—were the building blocks of the Military-Industrial Complex. As the dynamic relationships among them evolved over the course of the Cold War, four historical currents worked beneath the surface. *Technological determinism* seized the public imagination in the 1960s, beginning with Jacques Ellul's best-selling *The Technological Society*.[10] As the concept was teased out by other scholars in the ensuing decades, technological determinism came to have two meanings. First, technology appeared to be shaping, in some ways determining, the course of history. Second, some technologies seemed to evolve autonomously, to follow trajectories of their own making. Both versions of the term implied a loss of human agency, an inability of people to control their machines or determine their own future.[11] C. P. Snow, an astute observer of the human predicament, is reliably reported to have said in 1960 that thermonuclear war was a "mathematical certainty" within a decade if the world did not drastically reduce its stockpiles of nuclear weapons.[12] When that possibility was projected onto the nuclear arms race, it seemed that science and technology were moving the human community to Armageddon and the only brake on events was "mutual assured destruction." Eisenhower's warning moved from concerning to frightening.

A second and related current in the Cold War raised the prospect of *economic impact*. Eisenhower did not think that the United States and the Soviet Union would come to blows, but he did fear that the arms race could bankrupt both countries. That view saw arms expenditures as a drag on the economy, drawing talent and resources away from more productive investments such as education, infrastructure, and consumer goods. Others argued that defense spending actually stimulated the American economy, spinning off civilian applications, developing dual-use technologies such as radar and airplanes, and providing jobs in high-tech industries. This debate about the economic impact of the MIC stretched unresolved through the Cold War and beyond.[13]

Controversies over *statism* roiled US politics for two centuries and undergird the Cold War debate over the rise of a National Security State (NSS).[14] Did the perpetuation of a large standing military establishment, with its claims on a major portion of the federal budget, condemn the United States to an ever-larger national government and a centralization of power in Washington? Or would American anti-statist forces—checks and balances, interest-group politics, ideological diversity, and so on—prevent militarization? This question, too, survived the Cold War without resolution and remained controversial in the ensuing decades.[15]

Finally, the question of *collusion* hangs over the Military-Industrial Complex. Most interpretations of the complex echo the opprobrium that Eisenhower introduced in his farewell address, even while expanding the list of participants beyond just "industry." Different scholars have spoken of Congress as playing a role—or universities, or the media, or Hollywood. But no consensus on collusion ever arose. Did these individuals and institutions constitute a conspiracy, actively cooperating to shape national policy, raise defense spending, and possibly even militarize the United States? No doubt, instances of such collusion can be found in the historical record. Certainly, parties could independently place their own interests over the national interest. Industry did it all the time, as did many individual legislators. Countless players *mistook* their individual or parochial interests for the national interest. But did they consistently collude with the military, or did they simply share a common goal? And how were legislators to vote when the perceived national interest failed to coincide with the parochial interests of their state and district constituents?[16]

Opinions vary, but few students of the subject argue that collusion was essential, or even pervasive. Indeed, competition might have had the opposite effect, dispersing political influence among participants and disrupting collusion. The literature suggests instead that a culture of cooperation—sometimes mediated by the adversarial dynamics of principal-client relationships—often led many Cold Warriors to make common cause. Think of baseball teams, for example, engaged in a zero-sum rivalry with other teams and yet united with them in the common cause of promoting their sport and their league. Collusion and competition need not be mutually exclusive.

These five institutions—the state, strategy, the military, industry, and contracts—and four themes—technological determinism, economic impact, statism, and collusion—run throughout the narrative that follows. They are the ingredients of the MIC story. But they are not the story. The Military-Industrial Complex evolved over time, through the Cold War and into the decades beyond. That evolution was shaped by the dynamic interactions of these categories of analysis. In the text that follows, the story will be framed in a series of sketches, each one exploring how the ingredients interacted with one another and with their historical context. Part I presents the Cold War story. Chapter 1 provides a narrative overview of the MIC in the Cold War. The next five chapters update the thematic analysis of the Cold War MIC. Chapter 7 adds a brief analysis of US arms transfers during the Cold War.

Part II of the book tracks the way in which events have evolved since 1991. Chapter 8 explores the first decade after the Cold War. In a new, unipolar world, the United States enjoyed a hegemonic dominance of world economic, political, and military power. It attempted to inaugurate a "new world order." Chapter 9 examines the Global War on Terrorism following the attacks of September 11, 2001 (9/11). The ensuing wars in Iraq and Afghanistan drew the United States into a quagmire reminiscent of Vietnam. As in that tragic precursor, the United States military found itself ill-prepared and ill-equipped for the challenges it faced. Chapter 10 focuses on the second decade of the twenty-first century, when the United States sought a new grand strategy and a new arsenal to face the continuing threat of terrorism and the possibility of a new peer rival. The conclusion in chapter 11 explores whether the Military-Industrial Complex remains the country's institution of choice to arm and equip its national security forces.

The Cold War MIC, 1950–1991

Chapter One

Defining the Complex

T wo great ironies surround Dwight Eisenhower's warning about
the Military-Industrial Complex. First, Eisenhower had long
been an advocate of closer cooperation between the American
military and industry. As a young lieutenant he had observed
firsthand the lack of military-industrial cooperation in World
War I. Already a world-class industrial power by 1917, the United
States was slow to mobilize industry to serve its war effort. The
government spent more than $1 billion for aircraft in World War
I but only 960 planes reached the front, not a single fighter
among them.[1] Similarly, the American emergency shipbuilding
program of World War I spent $3 billion, paid twice as much per
ship as the British, and did not deliver a single vessel until the
last year of the war.[2] Postwar congressional hearings on the
"Merchants of Death" fed a perception that American industry
had profited from the war without contributing proportionally.[3]

Nothing between the wars altered the military's belief that
the next war would be less forgiving. World War I, the first "total
war" in history, had been a conflict pitting the total resources of
the state against those of its enemies. It had been a war of industrial production, won by the alliance that was able to field the

most combatants and provide them with the arms, equipment, food, and fuel necessary to fight in the machine age. Eisenhower understood that the next war would look the same. In a 1928 paper at the Army War College and again in a 1932 paper at the Army Industrial College, Eisenhower called for closer cooperation in peacetime between industry and the military, so that when war came they would be prepared to work in harmony.[4] The military had to communicate its needs to industry; industry had to meet those needs on short notice.[5]

What, then, caused Eisenhower in 1961 to warn against the very collaboration he had once promoted? World War II. Like World War I, this was a war of industrial production. The United States was not only the arsenal of democracy; it was the factory, the breadbasket, the warehouse, and the delivery service. Its national income rose 63% between 1939 and 1945, while the net income of the rest of the world increased not at all. Furthermore, it pioneered a new kind of military competition. This was the first war in history in which the weapons in play at the end of the conflict differed significantly from those at the beginning. Jet aircraft, guided missiles, radar, and the atomic bomb were just the most familiar of the new technologies. The quality of a nation's technological output rivaled quantity in the outcome of the war, fostering the perception among military observers that in the next war, quality might actually count for more than quantity. Perhaps World War III would be won by the best weapons, not the most weapons.

This perception helped to transform the military's relationship with technology. Throughout most of recorded history, commanders had resisted technological change, preferring the proven and familiar to the new and untested.[6] The introduction of gunpowder alienated whole generations of mounted knights, who foresaw an end to valor in war. Underwater warfare was dismissed as dishonorable and ineffective until it proved its potential in the American Civil War.[7] As recently as World War I, the US Navy had opposed new, indirect gunfire techniques, forcing reformer William Sims to take his case directly to President Theodore Roosevelt.[8] The British Army was so stodgy in World War I that Winston Churchill, as First Lord of the Admiralty and later as Minister of Munitions, sponsored research on an armored land vehicle within the naval establishment; its code name "tank" was designed to make the Germans think it was a water-storage vessel. Even the mundane machine gun, avatar of twentieth-century infantry combat, was opposed at first because of its profligate consumption of ammunition.[9] By World War II, US Army aviators had grown to appreciate the importance of qualitative advantage, but tankers had not.[10]

Following World War II, the military reversed this historic pattern and rushed to embrace new technology. In the United States, this conversion was overdetermined. World War II had demonstrated both what American scientists and engineers could do and what the Germans might have done had the war lasted longer. The demobilization immediately following World War II convinced the military services that Americans would not tolerate a large, standing armed force; the services would have to match the sheer numbers of the Soviet military establishment with technology. In time, they would come to call technology a "force multiplier," a factor that magnified the military effectiveness of a smaller fighting force. In the same vein, automated and sophisticated weaponry promised to improve the survivability of Americans in combat, minimizing the casualties that democracies abhor. Finally, the services concluded from a proposed National Research Establishment that if they did not set the agenda for military science and technology with "demand-pull," the scientists and engineers might design the coming battlefield with "technology-push"—a kind of technological determinism.[11] The technological enthusiasm that consequently swept the American military in the 1940s and 1950s, the impulse to assert agency over the armaments of the future, contributed greatly to Eisenhower's alarm.

The second great irony of Eisenhower's warning is semiotic. The binary term he introduced into popular discourse, "military-industrial complex," found a resonance out of proportion to its descriptive value or its historical precision. Privately, and more accurately, Eisenhower called the phenomenon the "delta of power," including Congress in a triangular nexus with the armed services and industry.[12] Collaboration among these three powerful institutions pressured him relentlessly to expand the armed forces and their budgets. Particularly galling to Eisenhower was the mindless pursuit of new and often redundant weapons systems, devices through which the services contended with each other for roles and missions and laid the groundwork for increased budgets and force levels to match their new equipment. At its worst, what Eisenhower labeled the Military-Industrial Complex also shaped foreign policy, basing its call for large budgets and more arms on an exaggerated representation of the military threat posed by the Soviet Union. Virtually all high-ranking military officers, most defense-industry executives, and many legislators with defense contractors in their states or districts formed a caucus of national security hawks.

Others have argued that Eisenhower should have included academia in the complex, for the network clearly extended its tentacles into the country's research universities.[13] But Eisenhower chose instead to issue a separate warning about "a scientific-technological elite." This language echoes the probable source of Eisenhower's warning, sociologist C. Wright Mills's influential 1956 book, *The Power Elite*.[14] Mills warned against the concentration of power in the hands of corporate, military, and political leaders. Though Eisenhower himself was as comfortable with scientists and engineers as he was with corporate executives, he nonetheless perceived that the "technological revolution during recent decades" was forming a new locus of power in America, centered on the development of military arms and equipment.[15]

Eisenhower's choice of the term "military-industrial complex" has piqued the curiosity of scholars for more than half a century. Historian Delores E. Janiewski has drawn attention to the National Military-Industrial Conference (1955–1961), which Eisenhower kept at arm's length, declining many invitations to address the group.[16] This ultra-conservative lobby and its successor, the American Security Council, were manifestations of the problem that Eisenhower chose to target in his farewell address. Some observers have argued that Eisenhower should have called out a "military-industrial-congressional complex." This formulation would have added a third side to Ike's "delta of power," what Gordon Adams, the dean of Washington budgeteers, called "the iron triangle."[17] The president's brother Milton, who helped with

FIGURE 1. Dwight D. Eisenhower (*right*), then president of Columbia University, is pictured visiting his younger brother Milton, then president of Pennsylvania State College (later University) in 1950. A decade later, as US president, Dwight Eisenhower relied on the counsel and assistance of Milton, then president of the Johns Hopkins University in Baltimore, Maryland, when preparing his farewell address. *Penn State University Archives.*

the speech, later reported that Ike himself actually inserted "military-industrial-congressional complex" in a draft of the address, but later removed it (figure 1). "It was more than enough to take on the military and private industry," Milton reported his brother saying. "I couldn't take on the Congress as well."[18]

Though Eisenhower's chosen label, "military-industrial complex," had little impact at the time, it assumed a prominent place in the national debate over the Vietnam War in the late 1960s and 1970s. Imitators invoked it as Eisenhower did, pejoratively.[19] The terms "military" and "industry" have complicated and ambiguous meanings deeply rooted in American history. While the country's security and well-being have depended upon its armed forces and industry, both have also been seen as potentially threatening to the country's core values. Americans inherited from their British forbearers a distrust of standing armies. Colonists in North America watched the motherland go through a wrenching civil war in the middle of the seventeenth century; citizens in their homeland suffered a suspension of traditional rights and privileges during the Cromwell interregnum. When it came time for the Americans to write their own constitution in the eighteenth century, they adopted constraints on military power similar to those that the British had invented in the Mutiny Acts of 1689. In addition to an implied preeminence of civilian over military power, the constitution provided that the army had to be funded annually by Congress. No American Cromwell would be able to lead a new model army in the United States without annual congressional assent.

American distrust of industry arose later. In what Mark Twain called "the Gilded Age," robber barons of the late nineteenth century demonstrated to Americans the dangers of unbridled free enterprise.[20] Technological marvels such as United States Steel, Standard Oil, and the great railroad systems of William H. Vanderbilt and J. P. Morgan stoked the American industrial revolution, but also provoked the Sherman Anti-Trust Act, the Populist and Progressive Movements, and a popular faith in state regulation of industry. Henry Ford and Thomas Edison became the new model industrialists in America, geniuses whose economic values and entrepreneurial styles appeared to be more in tune with the democratic principles underlying American consensus. Still, the reforms of the Progressive era failed to arrest the growth of corporate power in the United States. By the time of the Eisenhower administration, Charles E. Wilson, president of General Motors and nominee to be secretary of defense, could assert without embarrassment that the interests of General Motors and the United States were the same.[21] If the quintessential American industry could harbor such elitist views, then surely C. Wright Mills was right about the emerging "power elite." Industry, like the military, needed close scrutiny.

By lumping these two realms together under the heading of "complex," Eisenhower imposed a sinister and worrisome overtone to an already troubling association. The "military-industrial complex" smacked of conspiracy, and the "scientific-technological elite" did nothing to dispel the ominous implications of Eisenhower's warning.

Six decades later, the opprobrium implied in Eisenhower's language still hangs over the relationship between war, technology, and the state. Viewed more empathetically, the Military-Industrial Complex was a benign and patriotic alliance between the defense industry and the Department of Defense to ensure America's security in the face of a perceived existential threat. Military goals ranged across a broad spectrum. At one end, a bureaucratic imperative drove each service to expand its roles and missions, and with them its budget. At the other end of the spectrum, the Soviet Union challenged the armed forces of the United States to protect the country and its interests abroad against nuclear and conventional threats. Industry's goals ranged from profits on the one hand to public approbation on the other; defense contractors wanted to be both wealthy and reputable. Military and industrial goals converged in the Cold War task of building an arsenal equal to the perceived challenge posed by the Soviet Union.

America's Military-Industrial Complex in the Cold War was by no means the first instance in history when war and technology found themselves in symbiosis. Richard Cowen, for example, has discerned a military-industrial complex in the iron-producing Weald region south of London in the sixteenth century.[22] Paul A. Koistinen and other scholars have found precursors of Eisenhower's Military-Industrial Complex earlier in American history.[23] Indeed, the term has been imposed anachronistically on countless previous examples of collaboration between the military and industry.[24]

The first scholar to discern a precise historical antecedent of Eisenhower's Military-Industrial Complex was William H. McNeill. In his 1982 book *The Pursuit of Power*, McNeill construed the relationship between the British Admiralty and the arms industry in the Anglo-German naval race at the turn of the twentieth century as being analogous to the American version half a century later.[25] Naval officers and industrialists cooperated to shape British policy. They agreed to interpret German naval expansion as a threat to British security and to persuade the British people and their representatives in parliament that the threat required dramatically increased naval spending. Their campaign fueled the naval arms race and established a model for coordinated political action. The navy got more and better ships, and industry got contracts to build and equip those ships, all in the name of national security.

Eisenhower's Military-Industrial Complex differed from McNeill's and from the others discerned by recent historians in several important ways. First was its enormous scale and long duration. From 1950 to the present, the United States has sustained a permanent military mobilization in peacetime. That establishment shrank after the Cold War ended in 1991, but it did not experience anything like demobilization. No precedent for this establishment existed in all of American history. From the end of the War of 1812 until well after World War II, geography and politics had combined to provide America with free security.[26] Between 1945 and 1950, however, the administration of Harry S. Truman (1945–1953) concluded that the Soviet Union posed a real threat to world stability and thus to the security of the United States.[27] Truman vowed in 1947 to contain communist expansion. His secretary of state, General George C. Marshall, offered all nations material assistance in rebuilding their economies. Early in 1950, President Truman signed National Security Council Directive 68 (NSC 68), a landmark document that essentially committed the United States to permanent military mobilization in peacetime. The subsequent transition to a permanent war footing would have been more salient in the public consciousness had it not been masked by the Korean War. The United States military budget shot up during the war and never returned to pre-war levels.

The importance of technology in the ensuing Cold War became manifest in the 1950s. The Soviet Union exploded its own atomic weapon in 1949 to match the American achievement of World War II. In response, the United States embarked on development of a thermonuclear bomb, a fusion weapon orders of magnitude more powerful than the fission bombs of Hiroshima and Nagasaki. The defining feature of these weapons, however, was not their power but their cost. They were far cheaper than fission bombs in explosive power per dollar—providing "more bang for the buck" in Cold War parlance. These weapons allowed the two superpowers to build arsenals bristling with tens of thousands of warheads, enough destructive power to eliminate each other as functioning political entities and to irradiate their neighbors throughout the northern hemisphere. The United States tested its first hydrogen bomb in 1952; the Soviets approximated the feat just ten months later.

Second in importance only to the weapons themselves were their delivery systems. A perceived "bomber gap" of 1955 shook American confidence in the Eisenhower administration by suggesting that the Soviets had developed a fleet of bombers capable of reaching the United States. More alarming still, and far more threatening in actuality, was the launch of Sputnik I two years later. If the Soviet Union had the technology to place a satellite in orbit, then it could place a nuclear weapon on New York. Furthermore, while air defenses offered at least partial protection from bomber attack, no known technology could intercept an intercontinental ballistic missile (ICBM). The Soviet Union, hitherto seen as a backward and underdeveloped society, had produced two paired technologies that left the United States vulnerable to devastating attack.

In the wake of Sputnik, public fear bordering on paranoia, already manifest in the McCarthy hysteria of the early 1950s, broke over the Eisenhower administration, compounding the already relentless upward pressure on the military budget. The Military-Industrial Complex pushed public policy in the direction of more and better weapons, expansion of roles and missions, and mobilization of the civilian economy in the service of the state. The army and the air force launched parallel development programs to field intermediate-range ballistic missiles.[28] The navy and the air force raced each other to develop solid-fuel ballistic missiles, one to arm strategic submarines and the other to enhance the readiness of land-based ICBMs.[29] All the services proposed space activities to ensure for themselves a role in this new realm of human activity. While the United States continued to fight the Cold War with diplomacy, propaganda, foreign aid, and other traditional tools of international relations, the race for new and better weapons defined the East-West struggle.

As the services competed among themselves for roles and missions, they also attempted to harness significant shares of America's research-and-development talent and infrastructure. They continued some in-house capabilities, such as the Naval Research Laboratory and the Army's Redstone Arsenal. They sustained some World War II laboratories such as the Radiation Laboratory at the Massachusetts Institute of Technology (MIT) and the Johns Hopkins Applied Physics Laboratory. They supported the development of new think tanks such as the RAND (Research ANd Development) Corporation and the Institute for Defense Analyses, which depended primarily on military contracts. And they contracted with individual corporations, universities, and independent research organizations for "directed research" on subjects of interest to the military. From all of these sources, the services sought primarily "applied research," investigations directed toward the solution of specific problems. In general, they were expected to refrain from "basic" or "pure" research, designed to advance scientific understanding without necessarily serving the programmatic needs of any service or project. By government fiat, such basic research was the province of the National Science Foundation, the National Institutes of Health, and the country's universities and private institutions and foundations. All of these military research activities were overseen, abetted, and to some extent managed by the secretary of defense. His agents included the Advanced Research Projects Agency (ARPA) and his director of defense research and engineering, both created in response to Sputnik. Also responding to Sputnik was the National Defense Education Act, which provided support for graduate students pursuing advanced degrees in science and engineering.

Defense spending became a barometer of the nation's commitment to the highest quality arsenal in the world. The key was innovation, and defense spending on R&D was taken to be a measure of how the United States was doing. Of course, the real test could not be run until the American arsenal engaged Soviet arms and equipment in combat. But the development of a new weapon system could span five to ten years—sometimes more— from authorization to entering service. In this prelude to war, spending on research and development served as a metric of preparedness. The military funded more research and development in these years than any other segment of American society.

Nor did the race restrict itself to traditional military realms. In a war pitting the total resources of the state against its enemies, all agencies of government and all walks of life were subject to enlistment. Intelligence traditionally served both military and civilian purposes. President Truman created a Central Intelligence Agency (CIA) in 1947 to coordinate information gathering and analysis. The services, in turn, enlarged their own intelligence organizations and in time created a Defense Intelligence Agency. Atomic energy found itself consigned to a succession of agencies—the Atomic Energy Commission, the Energy Research and Development Administration, and the Department of Energy —all of which struggled to keep the production of nuclear weapons from driving their agendas. In the wake of Sputnik, Eisenhower empaneled a President's Science Advisory Committee, created the Advanced Research Projects Agency (ARPA, later DARPA),[30] and reorganized the Department of Defense—all attempts to manage the military's voracious appetite for new technology. The president also created the National Aeronautics and Space Administration (NASA, 1958) to prevent the heavens from becoming a new military arena. He could not, however, keep NASA from fighting the Cold War by other means.[31] Nor could he finally keep the military services from infiltrating space. He did, however, quash the army's proposal for Project Horizon, to establish a base on the moon for the simple reason that armies always "take the high ground."[32]

But he failed to prevent the services from using space for weather forecasting, strategic targeting, communications, and, of course, intelligence. Announced military spending on space exceeded civilian spending during the Cold War except for the two decades from 1961 through 1981, when the Apollo moon-landing program and development of the space shuttle swelled NASA's coffers.[33] But in those years and later, the classified budget for national security activities in space, including reconnaissance and signals intelligence, was hidden in the Department of Defense (DoD) budget. As Marcia Smith of the Congressional Research Service noted in 2004, "DOD sometimes releases only partial information [on defense spending in space] or will release without explanation new figures for prior years that are quite different from what was previously reported."[34]

The contest over defense spending persisted throughout the Cold War under the banner "How Much Is Enough"?[35] Could the country afford to place an economic limit on its own security? Eisenhower argued strenuously that it could, indeed it must. He conceived a "great equation" that balanced military preparedness against sustainable economic growth.[36] He expected the Cold War to be fought over "the long haul"; the winner would have to provide adequate security without bankrupting itself.[37] This calculus helps to explain the defense expenditures of the Cold War and especially the nation's investment in research and development.

The defense industry evolved to match government spending. In 1958, 30 of the top 50 companies on the Fortune 500 list of the largest industrial corporations also appeared on the list of the top-100 defense contractors.[38] Government contracts to universities—most came from the military—vaulted those institutions to national prominence as well. MIT and Johns Hopkins University, for example, held perennial positions on the list of leading defense contractors. Critics came to call it "the contract state."[39] Economist Seymour Melman warned that the United States, "if it is saved from nuclear war, will surely become the guardian of a garrison-like society dominated by the Pentagon and its state-management."[40]

Eisenhower's "delta of power" became an arena in which potent political and economic forces contended for defense contracts. Would-be contractors lobbied their legislative representatives to pressure the DoD to favor their bids. Legislators sought defense contracts on their own initiative, both to bring dollars to their district and to encourage campaign contributions from constituent contractors. The military services curried favor with legislative allies by placing contracts in their states and districts. Finally, the services realized that distribution of contracts and subcontracts across a broad array of congressional districts provided political insulation against cancellation of their programs. The "pork barrel" of nineteenth-century American politics turned into a powerful, sophisticated, complex infrastructure that presidents from Eisenhower to Clinton often found irresistible.[41]

The exact relationship between the locations of defense contractors and the voting behavior of their representatives and senators, however, remain unclear. Anecdotal evidence has suggested a strong causal effect. Not for nothing was Washington state's Henry "Scoop" Jackson known as the "Senator from Boeing," but early scholarship failed to confirm the expected correlations.[42] Two related factors seemed to be at work. First, the geographical distribution of defense contractors and subcontractors shifted dramatically during and after World War II. The hub of American industry moved from a heartland stretching eastward from the western Great Lakes—the so-called "rustbelt"—to a peripheral band stretching down the Atlantic and Pacific coasts and along the southern border—what Ann Markusen and her colleagues called the "gunbelt."[43] This "remapping" of the defense industrial base (DIB) certainly changed congressional voting patterns on defense spending, but no simple formulation explains the result. Various scholars have found multiple variables at work. They include strategy, cost, ideology, congressional committee assignments, and the relative economic impact of defense industries within their districts and states.[44]

Collectively, these studies leave unresolved the direction of cause and effect. Are defense contractors drawn to locations viewed as ideologically and politically congenial, or do regions adapt to the needs of these contractors as their economic impact grows? In any event, few of the studies challenge the correlation often found between the location of defense contractors and the votes of their senators and congressmen.

President Jimmy Carter, a former naval officer and a reform candidate for the presidency, discovered the power of the Military-Industrial Complex when he tried to cancel the B-1 bomber. Carter inherited from his predecessors a long-standing program to build a successor to the B-52, the workhorse of America's strategic bomber force and a mainstay of the air campaign in Vietnam. First proposed to President Eisenhower in 1957 as the B-70 *Valkyrie*, the replacement airplane was repeatedly deferred or rejected, either because its costs or capabilities were suspect or because the B-52 was still adequate. But Phoenix-like, the B-70 rose from the ashes, reinvented by the Military-Industrial Complex as a cruise-missile carrier, a "Long Range Combat Aircraft," and finally a "Strategic Weapons Launcher." President Carter thought he had terminated the program for good, but his successor, Ronald Reagan, especially sensitive to the aerospace industry in his home state of California, revived the airplane (now called the B-1) and put it into production in 1986. The last of the initial order of 100 planes came off the assembly line in 1988, just in time for the end of the Cold War. At $280 million apiece, more than ten times the projected price of the B-70 proposed to Eisenhower (three times the price in constant dollars), the aircraft never approached the performance capabilities promised for it.[45] Furthermore, it would have been obsolescent at birth had not its successor, the B-2 stealth bomber, been even later in development and far more over cost, rising finally to more than $2 billion per plane. Like the B-52 and the B-2, all designed to deliver nuclear weapons to strategic targets, the B-1 lives on as a battlefield bomber using smart ordnance to attack conventional military targets on land and sea.

Stories such as this one illustrate a characteristic of Cold War technology: obsolescence. US weapons systems often competed with themselves. Instead of measuring US technology against the capabilities of the Soviet Union or other potential adversaries, the Department of Defense often benchmarked its needs against an advancing wave of theoretical capability. The test of a weapon system was not its parity with the weapon systems of enemies or potential enemies but rather parity with the next generation of weapon systems that industry or the military services could envision. The United States had to develop the next generation, because it could not run the risk that an adversary might steal a march. The system drove itself relentlessly to build weapons not because they were needed but because they were possible. This suggested a kind of "technological determinism"—or perhaps more specifically a self-perpetuating "autonomous technology."

The B-1, which journalist Nick Kotz called the "born-again bomber," seemed to have a life of its own. Once proposed, it proved impervious to politics, economics, failures of performance, or even the waning of the Cold War. But it was not the technology that persisted and prevailed; it was the complex. The web of war, politics, economics, and technology formed a cocoon that protected weapon systems against diminished threats, corrupt politics, bad economics, or failed technology. Weapon systems routinely entered production LOCUS: late, over cost, and under specifications. Promoters of military technology mastered the technique of "buying in" and bailing out, Washington jargon for low bids at the outset and inflated promises later, both sustained until the sunk costs in a project made it too embarrassing for Congress to cut it off.[46] The B-1 was exceptional for its longevity and resilience, but it was hardly unique.

Nor did the phenomenon limit itself to the armed services. The civilian space program, another instrument of Cold War competition with the Soviet Union, embraced some DoD practices while eschewing others. The Apollo program succeeded in accomplishing the moon mission on time and on budget through a combination of capable management and adequate funding. NASA initially requested all the money it needed for Apollo, instead of asking for what the political market would bear. Then it solved its technical problems by pursuing multiple possible solutions simultaneously, as the Manhattan Project had done during

World War II. NASA funded alternative developments for risky components and designed redundancy into the complete system. This wasted money but hedged against program failure or delay.

Even during Apollo, however, NASA's overall record for research and development slipped into the range experienced by the Department of Defense. Its space science projects from 1958 through March 1996 experienced average cost growth of 220% above estimates.[47] The space shuttle proved to be a particularly glaring example. When NASA proposed this successor to the Saturn launch vehicle, neither Congress nor the administration of Richard Nixon supported the generous cushion of funding that Apollo had enjoyed. Forced to design to budget, NASA finally proposed a shuttle that could not possibly perform as promised.[48] It was a classic case of buying-in. When shuttle development ground to a halt in the Carter administration, NASA relied on the air force to win additional funding from the president. Like the B-1, the shuttle came in late, over cost, and under specifications. Many of the same players and imperatives that drove the Military-Industrial Complex had produced another marvel of technology that the country had not wanted or requested.

Did the United States, therefore, become a militarized state during the Cold War?[49] Were elected civilian leaders still in control of government? Or was the complex acting autonomously to produce technology of its own choosing? In 1941, political scientist Harold Lasswell had warned about "the garrison state . . . a world in which the specialists on violence are the most powerful group in society."[50] This possibility suggested itself to Lasswell when he learned of the Japanese bombing of China in 1937. He wondered if the subjection of civilians to military attack, what he called the "socialization of danger," would not empower the military to permanently organize society for war. The specialist on violence, he suggested, might displace the businessman, the bureaucrat, and the politician atop "the power pyramid." C. Wright Mills's "power elite" might be militarized, paradoxically, by "specialists on violence [who] are more preoccupied with the skills and attitudes judged characteristic of nonviolence. We anticipate the merging of skills, starting from the traditional accoutrements of the professional soldier, moving toward the manager and promoter of large-scale civilian enterprise."[51]

Two trends converged in Lasswell's garrison state. On the one hand, a conventional militarization of society allowed uniformed officers and military considerations to gain purchase in the formulation of national policy. Those given to such fears could see the appointment of General George C. Marshall as secretary of state in the Truman administration as evidence of this trend, to say nothing of the election of General Eisenhower. On the other hand, civilian leaders elected and appointed to oversee the military might themselves become imbued with many of the values Lasswell dreaded. Secretary of State John Foster Dulles envisioned a moral crusade against godless communism and was prepared to take the United States to the brink of Armageddon in defense of "security."[52]

Lasswell's "merging of skills" reads like a prescription for selecting Charles E. Wilson as secretary of defense. His appointment by Dwight Eisenhower, the soldier turned university president and then US president, suggested to some that what was good for the defense department was good for America. Indeed, the military's infatuation with planned obsolescence came to resemble the automobile industry's embrace of tail fins and annual model changes. The militarization of the "power elite," the merging of skills foreseen by Lasswell, seemed complete when Wilson was succeeded as secretary of defense by Neil McElroy, president of Procter & Gamble Company. His appointment evoked the specter of weapon systems being sold like laundry soap. There was more than a little irony, then, in Eisenhower's warning against the Military-Industrial Complex, and even a "garrison state,"[53] when his administration seemed to embody the concept.

Perhaps the most salient characteristic of the Military-Industrial Complex in its heyday was the stunning contrast between its marvels of technological achievement and its wretched institutional excesses. The complex produced the arsenal that won the Cold War and contributed to the transformation of modern life. Reconnaissance satellites tracked the cars of Soviet officials on the streets of Moscow. The F-111 swing-wing attack bomber flew at two and a half times the speed of sound but landed slowly enough to put down on an aircraft carrier. Nuclear ballistic missile submarines eluded enemy detection for months on end while bristling with multi-warheaded missiles preprogrammed to deliver irresistible, cataclysmic retaliation against any attack on the vital interests of the United States. Ground-based ballistic missiles in the United States stood on constant alert to launch on warning still more retaliation with intercontinental accuracy measured in hundreds of meters. In the Gulf War of 1991, the global positioning system (GPS) of satellites and ground stations allowed American troops to navigate the pathless deserts of the Mesopotamian Valley with speed and accuracy that befuddled the indigenous soldiers. Then GPS was made available to the private sector for commercial and even personal applications. Smart weapons allowed American soldiers to stand out of harm's way and launch strikes at room-size targets while minimizing collateral damage. Stealth technology attenuated the power of radar, one of the decisive technologies of World War II, allowing American aircraft to fly with impunity through the enemy's electronic defenses. Ballistic missile defense defied the best efforts of the Military-Industrial Complex, but just the possibility that the United States might achieve it contributed to the collapse of the Soviet Union. In the end, the complex built up an arsenal of conventional weapons and capabilities almost as destructive and decisive as the nuclear arms with which the country began the Cold War.

But that same complex also bred political, social, and economic scandal. Defense contractors charged $435 for a hammer and $1,868 for a toilet seat cover.[54] And the military paid. The "missile mess" of the Eisenhower administration suggested armed services more interested in competing with one another than with the Soviet Union.[55] Security classification routinely masked information from the American public that was well known to the Soviet government. The nation stockpiled 32,000 nuclear warheads, enough to destroy the Soviet Union many times over. A revolving door shuffled defense and industry executives back and forth between conflicts of interest. Defense industries received corporate welfare and insulation from market forces. Projects routinely finished late, over cost, and under specifications. Some finished not at all. Many of Eisenhower's scientific-technical elite served the complex in the name of serving national security. "Defense intellectuals" and "beltway bandits" (companies in the vicinity of the beltway surrounding Washington, DC, that specialize in contracting for government) provided support services for the complex.

Furthermore, the contracting parties, the military services and the DIB, developed an unhealthy codependency in which industries behaved badly and the military monitored them poorly. Perennial contract reforms failed to tame industry lawyers. Government bailouts rescued defense contractors from their own mistakes. A revolving door moved industry foxes in and out of the henhouse of government positions overseeing contracts. Companies would regularly underbid lucrative contracts that could be ratcheted up later when the government had sunk too much investment in the project to let it fail. Often the government would prop up contractors just to maintain surplus capability within the DIB. Many critics saw these rescues as "corporate welfare." In their monopsonic/monopolistic relationship free from normal market forces, industry "gold-plated" products and systems, and the military services succumbed to "capability greed" when imagining the technological features their systems might include.[56] Contractors and subcontractors distributed themselves, or were selected, in states and districts with sympathetic legislators on Capitol Hill, who could oversee and fund military procurement; government whistle-blower Chuck Spinney called this "political engineering."[57] Too often this chicanery and collusion crossed the line into outright fraud and criminality.

The good news is that the Military-Industrial Complex won the Cold War. For better or worse, it produced the arsenal that sustained deterrence long enough for the democratic, free-market system to prevail over the authoritarian, command economy of the Soviet Union. The bad news is that the complex exacted a price. As Sovietologist George Kennan warned in 1946, the United States ran the risk in the Cold War of mimicking the enemy, of becoming what it fought against. If the United States did not become a garrison state, as had seemed possible during the worst nights of the McCarthy era, it nonetheless took on many of the aspects of a command economy. This is an economy driven not by market forces but by the imperatives of state policy. Technology was the focus of those imperatives.

This particular combination of forces in Cold War America made Eisenhower's Military-Industrial Complex unique. All industrialized states in the twentieth century have institutionalized some relationship between war and technology. Many states developed this relationship in the nineteenth century and even earlier. Only in the United States, however, did the concept of an MIC become part of the public debate and the historical event. Born in World War II, energized in the McCarthy era, labeled by Eisenhower, masked by the Korean War, exposed by the Vietnam War, and seemingly defused by the waning of the Cold War, the Military-Industrial Complex cut a swath through American history unmatched by the experience of any other nation. No country claimed as much for its military technology or achieved as much. No country worried more about the militarization of its institutions. No country was shaped as forcefully by the science and technology of war.

Chapter Two

Civil-Military Relations

The Military-Industrial Complex challenged civilian control of the military. The common lament running through Harold Lasswell's garrison state, C. Wright Mills's power elite, and President Eisenhower's farewell warning was abrogation of the country's constitutional and traditional commitment to civilian rule. Samuel P. Huntington's classic 1957 study, *The Soldier and the State*, argued that military professionalism would keep servicemen loyal to civilian rule through the crisis of the Cold War.[1] But Eisenhower worried nonetheless, not that the military was being civilianized, but rather that some officers embraced, even sought out, the alliance with industry and Congress that undergird the MIC.

Participants in the complex pushed consistently for higher levels of defense spending. In the first year of the Ronald Reagan administration, Senator John Tower, chairman of the Armed Services Committee, quickly called for annual defense budget increases of 9% to 13% over inflation.[2] The question "how much is enough" pitted fiscal prudence against patriotism and loyalty. To suggest that a weapon system was not needed bordered on treason. Refusal of a military request for funding not only placed the nation's servicemen at risk, it endangered the entire population. Estimates of Soviet capabilities portrayed an adversary ten feet tall. Calls for balance and moderation were dismissed as soft on communism.[3]

The struggle for civilian control of the MIC reached something of a nadir in the TFX project. In 1961, Defense Secretary Robert McNamara brought to the Pentagon his Ford Motor Company experience with rational management and econometric analysis, as well as his own entourage of "whiz kids." When they learned that both the air force and the navy sought funding for new-generation fighter-bombers, McNamara insisted that they combine their development programs to build a single aircraft for both services. Deaf to pleas that air force planes should not be designed to land on aircraft carriers, McNamara insisted that cost-benefit analysis favored joint development. He was determined to accomplish what Eisenhower's three secretaries of defense had failed to achieve, civilian control of the military.

Variable-sweep wings provided a technical solution for the fundamental incompatibility between the air force and navy specifications. If the aircraft's wings could be swept back for supersonic flight through enemy defenses and then swept forward for low-speed carrier landing, the same design could serve both masters. Coincidentally, a NASA researcher advanced a novel design concept to solve the knotty mechanical and aerodynamical problems of variable sweep, and General Dynamics set about building the TFX. Over the protests and lamentations of air force and naval officers, and the critical scrutiny of some congressmen and aerospace industrialists, the project produced the F-111 (figure 2). Characteristically and predictably, the plane came in

FIGURE 2. The F-111, shown here with its wings extended. Secretary of Defense Robert S. McNamara insisted that this aircraft could meet the needs of both the air force and the navy. This exercise of civilian control of the military proved disappointing. *US Air Force.*

LOCUS, but it worked. The navy version was canceled before going operational, but air force versions flew limited missions in Vietnam, a 1986 raid on Libya, and the 1990–1991 Gulf War before the plane was retired in 1996. McNamara's critics insist that his attempt to dictate joint development displayed a profound ignorance of navy and air force missions—to say nothing of an indifference to their pleas—but scholars have found that the navy applied its technological criteria to serve its own interests.[4] "The technical disagreement," says political scientist Robert Art, "really represented a struggle by the air force and the navy to keep their identities separate, distinct, and autonomous."[5]

The evolution of ballistic missile guidance demonstrates the ways in which the military-industrial imperative could actually transform US strategic policy, a realm of civilian authority. In the 1960s, the United States adopted counter-value targeting for its strategic nuclear weapons. If the Soviet Union attacked the United States or its vital interests, the US would retaliate by attacking Soviet infrastructure: its cities, industrial base, and economic resources. At the time, US intercontinental ballistic missiles had insufficient accuracy to target the Soviet Union's strategic forces, in other words, its missiles and bombers, which, in any case, might already be in flight to the US by the time retaliation was ordered. The threat of this counter-value retaliation led to the aptly named policy MAD—mutual assured destruction—which saw the superpowers through some of their most dangerous confrontations.

Impervious to this policy, the MIC worked through the 1960s and 1970s to improve missile accuracy. This campaign was driven not by the Soviet threat to US national security, but by interservice rivalry between the air force and the navy. Both services sought a share of the strategic nuclear mission. The navy initially found itself at a disadvantage, because missiles launched from its moving submarines were inherently less accurate than missiles launched from fixed silos on land. The navy, therefore, supported research on improved guidance. The air force followed suit, not because it served national interests but because it served the parochial interests of the air force and its contractors. In the end, both programs achieved marvels of technological sophistication, guidance systems that could direct ballistic missiles thousands of miles and land them within hundreds of meters of the aiming point. In fact, the guidance became so good that the United States unwittingly developed the capability to attack the Soviet Union's land-based strategic forces in a pre-emptive first strike. This had not been the national intent, nor was it in the national interest, but the MIC had produced the result on its own initiative. The capability alarmed the Soviet Union and set off the Soviet arms build-up of the 1970s that so exercised the Reagan administration and spawned yet another round in the arms race. Civilian policymakers watched helplessly as their chosen strategic policy was obviated by military research and development.[6] Cases like these fueled the angst over techno-logical determinism and autonomous technology.

The TFX and ballistic missile guidance are only the most notorious of the instances in which the MIC challenged tradi-tional civilian control of the military. Routine behavior such as the placing of defense contracts, the hyping of new projects to get Congress to buy in, and the coordinated exaggeration of the Soviet threat by both military and industry all lessened the au-thority of civilian policymakers and increased the influence of the military in decisions ranging from the federal budget and taxes to foreign policy and aid to education.

The military also undermined civilian control by creating a parallel infrastructure to challenge the counsel of civilian insti-tutions. The services created their own think tanks, such as RAND Corporation. They sustained pre–Cold War research es-tablishments such as the Naval Research Laboratory, which op-erated in the traditional civilian realm of basic research. They remained the primary sponsors of laboratories at research uni-versities, such as the Applied Physics Laboratory at Johns Hop-kins University. In some cases, they funded semi-independent spin-off institutions, such as MIT's Lincoln Laboratory and

Draper Laboratory. They set up their own advisory committees of civilian scientists and engineers, such as the Air Force Science Advisory Committee. The technical experts who found themselves in the military orbit provided excellent, usually independent, advice. But as the sponsor, the Department of Defense could always cancel contracts, terminate memberships, and restrict funding. By controlling support for these institutions and individuals, the military cultivated its own experts to challenge the normal order of civilian policy formulation.

The military also constrained civilian control by limiting the flow of information. Technical secrets were among the most closely guarded of the Cold War. Policy, plans, and intelligence were the normal stuff of classification, but technical information allowed your enemy to duplicate and defeat your weapons. Entirely new categories of security clearance—special access programs (SAPs) and sensitive compartmented information (SCI)—displaced "top secret" in the hierarchy of classification. When Robert Oppenheimer's security clearance was revoked in the McCarthy era because his opposition to fusion weapons suggested softness on communism, his career as a nuclear physicist ended. As the armed forces came to deem more and more technologies crucial to national security, the web of secrecy spread out from weapons systems into the civilian economy. The Export Administration Act of 1979 required the government "to restrict the export of goods and technology which [*sic*] would make a significant contribution to the military potential of any other country."[7] The following year, the DoD prepared its first "critical technologies list," which was, of course, classified. The unclassified version published in 1984 identified twenty critical "arrays of know-how," ranging from computer hardware, software, and networking to instrumentation, telecommunications, and "vehicular technology."[8] By 1991, the list had changed significantly and the focus had shifted to promoting these technologies while at the same time ensuring "protection of scientific and technological achievements from transfer to unauthorized parties."[9] The program called for the government to monitor the dissemination of "critical" information whether or not that knowledge had been produced with government funding.

These unprecedented restraints on technology transfer coincided with an enlargement of the scope of security classification. Executive Order 12356 of 1982 led to more than 19 million "classification decisions" in the 1984 fiscal year.[10] The scientific and technical communities complained of government infringement on free speech and intrusion into the academy, the laboratory, and the conference hall.[11] In Senator Daniel Patrick Moynihan's assessment, the United States had moved from "a culture of secrecy" in the McCarthy era to "the routinization of secrecy" through the remainder of the Cold War and beyond.[12] Librarian Herbert N. Foerstel estimated that the United States published about half the world's unclassified technical information in the 1970s, less than 25% by 1993.[13]

The identification of critical technologies and the assignment of security classification to nondefense research contributed to the blurring of distinctions between civilian and military. Beginning in 1991, a National Critical Technologies list subsumed the military version, reflecting "a highly interdependent economy with substantial overlap between civilian and defense applications."[14] Uniformed officers evolved from warriors to managers of violence.[15] The development of the Polaris missile submarine produced a management system known as PERT (Program Evaluation and Review Technique).[16] More effective at impressing congressmen than aiding management, the process nonetheless radiated an aura of rational development and spread to non-defense government agencies. Drawing on experience developing bombers in World War II, the air force created a comparable system, "concurrency," to accelerate development of the Atlas missile.[17] When NASA's Apollo program fell behind schedule in 1967 after a fatal fire during ground-testing of the crew capsule, the agency imported retired air force general Samuel Phillips to restore order. Phillips applied concurrency to Apollo under the rubric of "all-up testing."[18] It saved the Apollo program but eroded the perception of NASA as a civilian agency.

The blurring of distinctions between military and civilian technologies finally gave way to open pursuit of "dual-use technologies." These had both military and civilian applications. They are as old as roads, ships, wagons, and hunting weapons. Not until the Cold War, however, did the United States embrace them as a national goal. Computers offer the best example of a general-purpose technology developed with huge infusions of military funding. ENIAC, the first digital electronic computer in the US, was built to compute ballistic tables; it was then turned to calculations for nuclear weapons development. The IBM AN/ SFQ-7 pioneered networking for the SAGE continental air defense system and then went on to be the prototype for the SABRE commercial airline reservation system. Magnetic core memory appeared first on the Whirlwind computer, developed at MIT for the navy and the air force. The Defense Advanced Research Projects Agency for many years provided the chief source of support for artificial intelligence research in the US. It also led the development of computer networks, computer graphics, and massive parallel processing architecture.

At first, the military supported such research in order to develop technologies for direct military application. By the 1990s, however, the services found themselves migrating toward support of some commercial technologies. Breaking with the long pattern of military support for highly specialized technology custom designed for military applications, the defense establishment came to see that it could leverage its own budget by helping the market to develop products appropriate for both military and civilian applications. Defense procurement could then realize economies of scale, such as the rapid price drops experienced in consumer electronics. This philosophy allowed the US military to send its soldiers into the Gulf War of 1990–1991 with commercial-off-the-shelf (COTS) laptop computers. The military continued to develop technologies the market would not support, such as high-performance computers and exotic materials, but its gradual drift toward dual-use technologies adumbrated a trajectory that would accelerate after the Cold War.

Chapter Three

State and Industry

U ntil World War II, the United States developed many of its military technologies in its own arsenals, shipyards, and laboratories. Interchangeable parts, for example, were perfected in the Springfield, Massachusetts, armory. Eli Whitney had introduced them to the US government and received a contract to deliver 10,000 muskets during the war of 1812. Perfecting interchangeability, however, proved more difficult and more expensive than Whitney had envisioned. Like many Cold War contracts, Whitney's came in late, over cost, and under specifications (LOCUS). Interchangeable parts were brought to fruition only through painstaking and costly development in an arsenal insulated from market forces. Even there, it met resistance from an entrenched guild culture reluctant to change age-old patterns of handcraftsmanship.[1] From the arsenal, the concept of interchangeability spread to the private sector, through sewing machines, bicycles, and finally the automobile assembly line of Henry Ford.[2]

Of course, the military traditionally purchased from contractors as well—steel producers, commercial shipyards, munitions manufacturers—but it often looked to its own arsenals and laboratories for innovation and the establishment of standards. From naval ordnance at the Washington Naval Yard to aeronautical research at McCook Field (later Wright-Patterson Air Force Base) in Dayton, Ohio, to ship design at the David Taylor Model Basin outside Washington, DC, the military services supported their own research establishments and turned seldom to industry or universities for innovation.

All that changed in World War II. Though important work continued to flow from military R&D establishments during the war, a new locus of innovation arose to dominate the field. Rejecting the failed World War I model of commissioning scientists and engineers and bringing them onto active duty for the war emergency, the government created the National Defense Research Committee in 1939 and transformed it into the Office of Scientific Research and Development (OSRD) in 1941. Under the leadership of Vannevar Bush, former dean of engineering at the Massachusetts Institute of Technology and most recently the president of the Carnegie Institution in Washington, OSRD recruited scientists and engineers for the war effort but left them in place in their home institutions. Rather than inducting researchers into federal service, the government relied on contracts with individuals and institutions to harness technical talent and fund construction of the research infrastructure necessary at their institutions. The Radiation Laboratory at MIT, for example, became the nucleus of radar research in the United States, part of a network of sites working on more than 100 radar projects ranging from airborne anti-submarine radar to the proximity fuse, which allowed bombs and shells to explode when they came near their targets.

At the close of World War II, Vannevar Bush recommended to President Franklin Roosevelt a national research establishment to continue the wartime pattern of supporting the nation's scientists and engineers by contracting for research essential to national security, health, and economic development.[3] Bush's master plan foundered on congressional objection to entrusting scientists with the direction of these research institutes.[4] Instead, the government expanded the existing National Institutes of Health, created a National Science Foundation for basic research, and surrendered direction of military R&D to the newly created Department of Defense and its subordinate armed services. A proliferation of laboratories, advisory committees, and engineering centers sprang up to compete for funding. Interservice rivalry dominated the quest for strategies, roles and missions, and their attendant arms and equipment.

The arsenal system also survived, especially for those research projects requiring especially expensive facilities and equipment.[5] For example, the nuclear research and production facilities of World War II's Manhattan Project survived the war in only slightly altered form. Los Alamos Laboratory, where the Hiroshima and Nagasaki bombs were designed and built, continued weapons research in the Cold War under the management of the University of California. Sandia National Laboratories grew up next door, a commercial contractor responsible now for building the nation's nuclear weapons. University of California, Berkeley's wartime Radiation Laboratory was joined by a new Lawrence Livermore National Laboratory in 1952 to expand the range of weapons research being conducted under contract through the university.[6] The air force built an entire suite of wind

tunnels at Arnold Engineering Development Center in Tula-
homa, Tennessee, to conduct for itself the testing and innovation
previously done by the National Advisory Committee for Aeronautics.

In spite of the state-of-the-art equipment and lavish budgets
at these and other military laboratories, government research
facilities lost some of the luster they had enjoyed prior to World
War II. They were unable to overcome the long-standing perception that the best scientists and engineers gravitated to prestigious university and industry appointments, while second-class
researchers accepted bureaucratic positions with the government.
Arsenals, it was widely believed, were staffed by place-fillers and
mediocrities who embraced the security of civil-service tenure in
lieu of competing in the great marketplace of ideas and achievement. In spite of anecdotal evidence to the contrary, innovation
seemed to spring from universities and industry, while routine,
pedestrian results emerged from government laboratories.[7]

This perception fueled creation of a two-tier reward system,
in which the most distinguished scientists and engineers held
high-paying positions in academia and industry, supplementing
their salaries with lucrative government contracts and consul-
tancies, while lesser researchers worked at civil-service scale in
government laboratories. The power and pitfalls of this system
were nowhere more evident than in the rise of Ramo-Wooldridge
Corporation, a classic case study of the Military-Industrial Complex at its best and worst. Simon Ramo and Dean Wooldridge
were scientists-turned-engineers in the electronics division of
Hughes Aircraft Company. In 1953, Trevor Gardner, the special
assistant to the secretary of the air force for research and devel-

opment, visited the Hughes Aircraft Company plant in Culver City, California, and consulted with his old friend, Ramo. Gardner was impressed by Ramo and Wooldridge's "excellent leadership" of Hughes's Falcon air-to-air missile program. Following the visit, the two young engineers left Hughes to form their own consulting firm. The following year, Gardner awarded them a letter contract to provide "technical services" to a blue-ribbon Strategic Missiles Evaluation Group. The Group recommended to the air force a ballistic missile program. Ramo-Wooldridge received the contract to assume "technical responsibility" for that program. The resulting Atlas missile became America's first ICBM.[8]

Beginning with a small office and a handful of employees, Ramo-Wooldridge rapidly expanded in size and function, moving from management to manufacturing as well. By the time the company merged with Thompson Products Company of Cleveland to form TRW in 1958, the young engineers' initial investments of $6,750 had grown to $3,150,000 each ($28 million in 2019 dollars) and they still controlled the voting stock. The sole-source contracts awarded to Trevor Gardner's old friend had all the earmarks of a conflict of interest. The Atlas program succeeded, but at the price of high salaries and alarming profits for TRW.[9]

The facts of the Ramo-Wooldridge story are not in doubt; most of them came out during the course of a 1959 congressional investigation into the company's relationship with the air force. But what do they mean? Ramo-Wooldridge Corporation may be seen as a product of privilege and favoritism, exacting unwarranted profits from an indecent conflict of interest. But the Atlas missile may be seen as a technological marvel resulting from a unique alliance of free-enterprise capitalism and government sponsorship. Eschewing the socialist model of many other industrialized states, such as Japan and France, to say nothing of the command economy of the Soviet Union, the United States linked the public and private sectors in a cooperative scheme that retained strong aspects of both—the government's ability to make enormous capital investments in long-term projects benefiting the country, and industry's ability to mobilize the private sector for big undertakings. Historian of technology Thomas P. Hughes argued that such large technical systems occupy the apex of modern technological achievement.[10]

Despite the impressive results of this relationship between industry and the state, a number of significant shortcomings marred the overall record. Four may be singled out as particularly troubling and ambiguous: the "revolving door," the interference with market forces, the risk of corporate welfare, and the tolerance of corruption.

First, while it is true that the "revolving door" between public and private service allowed cross-fertilization of the two sectors, it also invited conflict of interest. Military officers retired and went to work for the companies whose contracts they had recently managed. Corporate executives took leave from their companies to enter government service in positions dealing with their former and future employers. Of course, changing places had much to recommend it. When Nora Simon and Alain Minc surveyed the computer industry for the French government in 1980, they concluded that the great advantage held by the United States in the world market resulted from the "thick fabric" of interwoven American institutions. Computer professionals from academia, industry, and government collaborated with one another and moved seamlessly among the different realms of computer activity. Simon and Minc believed that institutional rigidity in France precluded such healthy cross-fertilization there.[11] But the other side of this coin revealed collusion, corruption, and conflict of interest. An "old boy" network—in its heyday the MIC was predominantly male—swapped jobs, information, and confidences in the sure belief that what was good for the defense industry was good for America and vice versa. Admittedly, competition was intense, between companies, between services, and between educational institutions. But barriers between these realms were porous, allowing some actors to circumvent the rules.

Second, the military-industrial relationship also distorted market forces. Corporations, such as the Martin Company, limited themselves almost exclusively to military contracts. Such companies are more likely to give the military services whatever they demand, whether or not it is technologically sound, and less likely to develop new products to compete in the marketplace.[12] This condition heightens the impact of government specifications and lowers the impulse to innovate. On the other hand, military impact on the aerospace industry has also produced commercial spin-offs of great value. The Boeing 747, for example, commercialized the company's failed attempt to win a government contract for military cargo planes, just as the air force's C-135 was a converted Boeing 707. R&D contracts were one of the ways in which the government contributed to the rise of the American aerospace industry, the world leader throughout the Cold War.[13]

The government shaped defense-related markets in other

FIGURE 3. Marines and sailors board a Marine Osprey at Camp Buehring, Kuwait, on April 12, 2019. The "tiltrotor" aircraft's engine nacelles are positioned for a vertical takeoff. *https://www.defense.gov/observe/photo-gallery/.*

ways as well. Government standards, specifications, accounting practices, protocols, and record-keeping influenced the corporate style and worldview of defense contractors. On the one hand, the government might require levels of performance that the market would not support, as it did, for example, in anti-lock braking systems first developed for military aircraft and then transferred to the commercial automobile market. On the other hand, it might also promote refinements of technology more expensive than they are worth on the open market, as was the case with the V-22 Osprey (figure 3). Defense Secretary Dick Cheney told Congress in 1989 that the military simply could not afford these innovative, tilt-wing aircraft, then projected to cost $25 billion over the coming decade. Congress made DoD buy the aircraft anyhow.[14]

A third controversial aspect of the military-industrial relationship was corporate welfare. Believing that national security depended upon a wide and diverse defense industrial base (DIB), government officials distributed defense contracts so as to sustain multiple suppliers of critical products and services. Lockheed Aircraft received the contract for the C-5A transport over rival Boeing Corporation because Boeing had plenty of business at the time and Lockheed was struggling. Air force leadership overruled its own technical panel to make this award. Such tinkering with the free market made good sense militarily, but it flirted with the socialism that so many defenders of the MIC professed to abhor. In short, it turned many defense contractors into quasi-nationalized industries sustained for reasons of state policy, as opposed to market competitiveness. Seldom is this model credited with promoting innovation.

Fourth and finally, the intimate working relationship between the military services and the defense industry invited outright corruption. The extraordinary C-5A scandal illustrates this problem as well as any.[5] In 1965, Lockheed Aircraft received the $3 billion contract to produce an aircraft capable of take-off and landing at unpaved airstrips, cruising at 600 miles per hour, and lifting more than 200,000 pounds of cargo. By 1971, the cost of each airplane had swollen to almost three times the original bid. What really set the case apart, however, was the craven politics of the senators and congressmen whose states and districts profited from the contract. Representative Mendel Rivers (D, SC), whose re-election slogan was "Rivers Delivers," went so far as to threaten a colleague on the House floor with loss of military jobs in his home district. He proclaimed during hearings, "Regardless of what the plane costs, we need it, and we must have it."[16] The air force compounded the public embarrassment by insisting that nothing was amiss. The civil servant who brought the scandal to light, Berkeley Rice, was punished for his disclosures.

More than delays and cost overruns plagued the C-5A. It was also a technological disappointment. Upon delivery, the giant aircraft was found to have wing defects that limited cargo capacity to 100,000 pounds, less than half the prescribed load. Defective engine mounts precluded full opening of the throttles, eliminating flight at full speed and take-off and landing from unimproved runways. Poor landing gear kept the plane from putting down in a cross wind. Other failures further limited operational capability. Research and development on cutting-edge technology often produces delays and cost overruns, but if the process does not generate the expected technological advances, then the whole relationship between industry and the state falls under a cloud of suspicion.

Chapter Four

Among Government Agencies

W ithin the federal government, the Military-Industrial Complex spread its influence far beyond the confines of the Department of Defense. Most of the agencies that felt its impact operated in technical realms, but even diplomacy and the environment were affected. Foreign aid to developing nations often took the form of credits for purchases of US military arms and equipment. Allies such as Israel demanded state-of-the-art American weapons to fight their Arab neighbors, who were armed largely with Soviet weapons. American arms manufacturers sought licenses to sell their wares abroad, seeking economies of scale that would subsidize the defense industry. And all sales of US military technology gave the Department of State leverage with the purchasing countries, which would need spare parts, technical assistance, and training to maintain and operate the equipment. In short, the MIC produced a cornucopia of sought-after technology that the Department of State could allocate in pursuit of policy goals. Sometimes, however, the economic-industrial imperative came into conflict with avowed national policy and the State Department found itself mediating conflicts in which both sides were armed with American weapons.[1]

The Environmental Protection Agency (EPA) received from the MIC powerful tools of monitoring, regulation, and enforcement. Earth-resources satellites, for example, drew upon military rocket, satellite, microelectronic, and sensor technology to detect water and soil pollution and to monitor spring run-off from winter snows. But the EPA was also frustrated by national security exemptions to environmental regulations, ranging from exhaust emissions to noise control. During the Cold War, nuclear weapons facilities across the country are believed to have contaminated 475 billion gallons of ground water. The Department of Energy estimated in 1989 that it would spend $147 billion over 75 years to clean up 113 sites.[2]

Nowhere was the impact of the Military-Industrial Complex felt more profoundly than in the quasi-military world of intelligence. The arms and equipment needed by the United States were determined in part by the capabilities and intentions of the Soviet Union. The perverse logic of the Cold War convinced hard-liners in both the United States and the Soviet Union that those determinants could be reduced to one: capabilities. Each side felt compelled to assume that the other side might do anything it could do. Intentions were subjective; capabilities were matters of fact. Or so the argument went. Attention, therefore, focused on the force structure of the Soviet Union, with an eye to assuring that US force structure provided an adequate deterrent. If the Soviet Union developed a Mach 2 fighter plane, such as the MIG-21, then the United States had to have a Mach 2+ fighter, the F-4, or better yet, the Mach 2.5 F-15. If the Soviet Union developed an intercontinental ballistic missile with guidance accuracy of 100 meters, then the US had to harden its missile silos to withstand 2,000 pounds per square inch of overpressure. If Soviet sonar improved its sensitivity, then US nuclear submarines had to have propellers virtually free of noise-producing cavitation. The arms race of the Cold War was driven in large measure by the intelligence each side had about the other's capabilities. Seldom was it constrained by subjective analyses of how the enemy intended to use its arsenal. Furthermore, innovations, both offensive and defensive, raised the risk to one side or the other. The net danger ratcheted up with seemingly deterministic or autonomous logic.

This competition put a premium on the intelligence itself. Service roles and missions would rise and fall on the perception of enemy threat. If ground-based missiles and bombers were vulnerable to preemptive first strikes, this raised the stock of the navy's ballistic missile submarines. If Soviet armored forces posed a threat to Europe, then anti-tank weapons and close air support gained purchase in the United States. If the Soviets developed a workable anti-satellite missile, then a new and independent US space command might be required to enhance the growing military infrastructure in earth orbit.

For all these reasons, intelligence became politicized. The Central Intelligence Agency (CIA) found itself besieged. Individual services dissented from its "national intelligence estimates" and offered their own interpretations to the president and the National Security Council. The Department of Defense created its own Defense Intelligence Agency to both gather and interpret information in parallel with the CIA. Aircraft and then satellite reconnaissance became so important that a separate, independent agency, the National Reconnaissance Office (NRO), was spun off to conduct that business. It, too, became a stepchild of the MIC, relying on most of the same contractors and consultants, and weaving another "thick fabric" of interconnected institutions and personnel. Still another technological spin-off appeared in 1952 when the National Security Agency (NSA) set up shop apart from the CIA to intercept, decode, and analyze the world's communications. When the CIA first publicly disclosed the annual intelligence budget of the United States in 1997, it was revealed to be $26.6 billion, much of it invested in the development, maintenance, and exploitation of high technology.[3] Everything from supercomputers to sensing devices to artificial intelligence for machine translation and photo interpretation fed on the intelligence imperative.

Few civilian federal agencies have experienced as close and ambiguous relationship with the MIC as NASA. Intended by President Eisenhower to keep the military from capturing the national space mission and militarizing the heavens, NASA tried from the outset to distance itself from the Department of Defense. It stressed the priority of space science. It left reconnaissance to the military services. It invited international cooperation and openness. And it filtered its military relations through an innocuous and impotent Aeronautics and Astronautics Coordinating Board.

But it was always a tough sell. In the late 1950s and early 1960s, NASA flew its missions on military missiles converted to launch vehicles, its astronauts were military test pilots, and its satellites were indistinguishable from the reconnaissance spacecraft that were one of the most poorly kept secrets of the Cold War. Indeed, the first US satellite reconnaissance program, Corona, flew under the cover of being a NASA science project, Discoverer. When U-2 pilot Gary Powers was shot down on a spy mission over the Soviet Union on the eve of the 1960 summit conference, NASA Public Affairs Officer Walter Bonney insisted that Powers had strayed off course on a NASA weather mission. That cover story evaporated when Powers was captured and confessed. With it went much of NASA's credibility as an independent agency innocent of Cold War machinations.

Even as NASA grew in independence and stature during the Apollo program of the 1960s, its ties to the military remained visible. It absorbed the Army's Redstone Arsenal and renamed it in honor of General George C. Marshall. It continued to fly its test aircraft out of a facility spun off from Edwards Air Force Base in California, and it created its own space launch facility at Cape Canaveral, Florida, next to Patrick Air Force Base. When the Apollo program got into trouble, it recruited an air force general to turn things around with an air force management scheme. At the end of the Apollo program, when it was trying to sell the shuttle as a follow-on vehicle, it designed the orbiter to fly air force reconnaissance missions in order to get much-needed military support for the project. When the shuttle faced cancellation in the Carter administration, the air force was once more enlisted to save it. Though NASA moved toward civilian astronauts and its own family of launch vehicles, it never lost its close ties to the military. Nor was it immune to developing a relationship with the aerospace industry that became hard to distinguish from the military. Indeed, NASA even imitated some of the military's contracting techniques, not only technical schemes such as cost-plus-fixed-fee contracting but also political techniques such as buying-in and distributing subcontracts by congressional districts. Just as the space program was a continuation of the Cold War by other means, so, too, was NASA's relationship with the aerospace industry a continuation of the MIC by other means.[4]

Nuclear power provides another instance of a critical national technology with dual-use potential firmly rooted in the Military-Industrial Complex. Believing that the atom was too important to be left to the generals, the United States created the Atomic Energy Commission (AEC) in 1946. Successor to the wartime Manhattan Project, it assumed responsibility for both the military and the civilian development of this promising yet frightening technology. At the height of the Cold War, it concentrated on development of fission and then fusion weapons. In the 1960s, the AEC promoted and launched commercial nuclear power in the United States and oversaw a surge in plant construction known as the great bandwagon movement. Companies such as General Electric, which had pioneered military development of the atom in nuclear propulsion for naval vessels, converted that technology into turn-key, light-water-reactor plants that promised, in the words of Atomic Energy Commissioner Lewis L. Strauss, "power too cheap to meter." But that decision, to build a US commercial nuclear power industry on the light-water reactor developed for naval applications, contributed to subsequent problems. This was a classic instance of what economists call "path dependence," a technological outcome shaped by the trajectory of its development.[5]

The public, however, had difficulty dissociating the commercial reactor down the road from the mushroom cloud over Bikini

Atoll. "Nuclear fear," historian Spencer Weart's term, permeated popular culture and tainted the perception of commercial nuclear power.[6] Anti-nuclear activism of the late 1960s and early 1970s, augmented by the nascent environmental movement and concern for nuclear waste, derailed the great bandwagon movement and crippled the commercial nuclear power industry in the United States. While other countries such as France and Japan proceeded apace with the commercial development of nuclear power, the United States turned away from this technology for the remainder of the twentieth century. In no small measure, the reversal of course resulted from associations in the public mind between commercial nuclear power on the one hand and the bomb and the MIC on the other.

In an effort to regain civilian control of this technology, the US government experimented with other institutional arrangements in the 1970s and 1980s. The AEC was broken up in an attempt to separate its conflicting responsibilities for regulation and promotion of nuclear energy. The Nuclear Regulatory Commission thereafter oversaw civilian nuclear power but exercised little authority over military facilities and programs. The Energy Research and Development Administration took up promotion of national energy resources, including nuclear, but found itself absorbed by the Department of Energy (DoE) in 1977. Responsibility for producing the nation's nuclear arsenal migrated to the DoE, as did many veterans of the old AEC weapons programs. The power of civilian commissioners to control the weapons program remained a source of controversy throughout the Cold War and beyond. When the environmental abuses of the weapons laboratories were made public in the 1990s, it became clear that national security had trumped civilian governance in the AEC and its successor agencies. The impact of the Military-Industrial Complex on the executive branch of government reached far beyond the Department of Defense.

Chapter Five

The Scientific-Technical Community

P resident Eisenhower's warning about a "scientific-technological elite" shocked the scientific community. Eisenhower had
created a President's Science Advisory Committee in the wake of
Sputnik, and he developed a great respect and affection for the
scientists and engineers who served him in that capacity. What
troubled him was the prospect not only of the growing power of
such experts over public policy but also the related "prospect of
domination of the nation's scholars by Federal employment,
project allocations, and the power of money."[1]

In addition to the examples at MIT and the Johns Hopkins
University already mentioned (chapter 1), other universities and
research organizations rose to national prominence on a wave of
government funding. The University of Washington became the
second largest university defense contractor (behind MIT) by
expanding its World War II Applied Physics Laboratory. Lincoln
Laboratory, created by MIT in 1951 to develop an air defense
system for the air force, quickly expanded into other areas of
defense contracting. In 1958, it founded the nonprofit MITRE

Corporation to support the Air Force SAGE project, developing a system of early-warning radars. By 1986, Lincoln had spun off 48 companies, totaling $8.6 billion in annual sales and employing more than 100,000 people. Professor Charles Stark Draper's Instrumentation Laboratory, in MIT's Department of Aeronautical Engineering, rivaled Lincoln Laboratory by the early 1960s and dwarfed its parent department. By 1965, it had spun off 27 companies with sales totaling $14 million and employees numbering 900.[2] In 1989, three universities and two spin-offs were among the top 50 research-and-development contractors for the DoD: MIT (13), Johns Hopkins (17), the University of California, San Diego (43), MITRE Corporation (18), and Draper Laboratories (28).[3] Bolt, Beranek, and Newman, a Cambridge consulting firm specializing in acoustics and computers, became something of a halfway house for MIT faculty and students venturing into the technical marketplace. Stanford University was playing an equally decisive role in launching the Stanford Industrial Park and Silicon Valley.[4] At these centers of excellence, as they were called by the DoD's Advanced Research Projects Agency, and at other universities around the country, academia blended with the Military-Industrial Complex in ways that blurred the boundaries between them.

In one sense, this development was natural and positive. Academia, after all, made up one of the major threads of that "thick fabric" that Simon and Minc credited with American computer achievement. The issue that alarmed Eisenhower was proportion. Would the government, and especially the military, come to play too large a role in university policy, and would scientists and engineers establish unofficial rule over a Washington technocracy increasingly driven by technological imperatives? Many scholars examining the influence of military funding on science and technology perceived a subtle but profound skewing of research agendas.[5] The independence of university researchers in their laboratories succumbed to the reality of funding imperatives, especially in high-tech fields such as nuclear physics, aerodynamics, and supercomputing that required expensive research facilities.

Furthermore, the intrusion of defense dollars into university budgets influenced institutional policies in ways that spread beyond individual faculties and laboratories. The National Defense Education Act of 1958, one of the many national responses to Sputnik, provided scholarships and fellowships for students working in scientific and technical fields valued by the military. Faculties grew because many members taught reduced teaching loads, bought out of the classroom by research dollars from the DoD. As universities added teaching faculty and made additional research appointments funded by government contracts, support infrastructure grew up around them—offices, services, parking spaces, etc. New research facilities were built at government expense and added to the universities' physical plant, requiring maintenance and still more support staff. Only annual infusions of funds from the DoD could support this vast infrastructure. In 1984, 36% of MIT's engineering research budget came from the DoD, as did 71% of the funding for its Laboratory for Computer Science, 62% for the Artificial Intelligence Laboratory, and 40% for the Research Laboratory of Electronics.[6]

The military presence in the academic community entailed more than financial dependency. Secrecy visited the university campus as a condition of government largess. Most military research essential to national security was classified. This constraint breached academic norms of publication and open discussion. Most academics recognized the need for a certain amount of confidentially in their work; for many, priority of publication was the currency of professional advancement. Access to preliminary research results was often restricted, but publishing was always the goal. Research did academics no good until the results appeared in print, preferably in prestigious, peer-reviewed journals. Military work, however, might never see the light of day. The more critical it was to the military services, the less likely it would be cleared for publication. Scientists who accepted these constraints could find themselves cut off from the normal round of papers, conferences, and published literature. Instead, they communed with other defense contractors and shared their research results only with the services sponsoring their work.

During the Vietnam era, the issue of secrecy achieved an exceptional political salience on university campuses. At Stanford in 1969, for example, the dean of engineering announced that the school would no longer accept classified contracts. The academic senate went one better by effectively banning all classified research from campus. The following month the trustees announced that the university would divest itself of the Stanford Research Institute (SRI), which had been incorporated in 1946 to promote research and education at Stanford. By 1969, most of the SRI's funding was coming from the DoD.[7] These responses to student and faculty protests had serious consequences for those scientists and engineers whose regular research agendas relied heavily on defense dollars. Laboratories, research associates, and graduate students often depended on that funding. Without it, many faculty would have to change fields or change institutions. In the end, the political pressure subsided at Stanford, and the restrictions on classified work slackened. However, the fundamental incompatibility of secret research and academic ethos had been dramatically demonstrated. Similar struggles occurred on other campuses around the country.

As the pressures of the Cold War abated in the 1970s and 1980s, institutional prohibitions against defense funding subsided as well, even as a new and pernicious phenomenon crept onto campuses and research establishments around the country. The blurring of distinctions between military and civilian technologies, the embrace of dual-use technologies, and the growing overlap between military and economic security prompted the government to expand the scope of its classification protocols. The identification of military-critical technologies in the 1980s and the spread of this categorization to other branches of government coincided with the Reagan administration's crackdown on technology transfer abroad. After the worst instances of this government intrusion into academic life in the early 1980s, compromises were negotiated to balance the researcher's right to publish against the government's right to restrict the flow of information crucial to national security, but the experience suggests that the military/academic confrontations of the Vietnam era had not entirely dissipated.[8]

More troubling still to many civilian scientists and engineers were the moral implications of research in the service of the MIC. Even before the complex took shape, the atomic bomb alerted many scholars and academics to the potential militarization of their professions. Manhattan Project veterans founded the Federation of Atomic Scientists in 1945, now the Federation of American Scientists. The group monitors science, technology, and public policy. In 1946, veterans of the Manhattan Project also began publishing the *Bulletin of the Atomic Scientists*, with its ominous doomsday clock on the cover, warning of imminent nuclear cataclysm. In 1957, 22 scientists from 10 countries met in Pugwash, Nova Scotia, to discuss the threat posed by nuclear weapons. Since then, hundreds of Pugwash conferences, symposia, and workshops have brought together scholars and public figures to address the world's problems, most of them tied to military science and technology. In 1995, the Pugwash conferences and their president, Joseph Rotblat, shared the Nobel Peace Prize. The Union of Concerned Scientists was founded in 1969 by a group of students and faculty at MIT to advocate scientific research on social and environmental problems instead of military programs. Computer Professionals for Social Responsibility (CPSR) formed in the late 1970s and crusaded through much of the 1980s against the automation of warfare, especially the loss of human agency inherent in the increasing use of computers for command and control of weapons of mass destruction. Computers, they believed, lent themselves to a kind of technological determinism, in which machines might one day decide what Jonathan Schell called "the fate of the Earth."[9] CPSR advocated that its members refrain from working on DoD projects or accepting military funding.

A claim of guild privilege colored all such movements, but usually they were couched in moral terms.[10] Scientists and engineers, they asserted, had a contract with society to use their special talents for the good of humanity. Science and technology should meliorate the human condition, not produce instruments of coercion and destruction. While many citizens, including scientists and engineers, saw military arms and equipment as guarantors of national security and bulwarks against communist aggression, the protestors focused on the excesses of the MIC. They believed that it posed a greater threat to the nation than the remote Soviets and their fellow travelers in Europe, Asia, Africa, and Latin America. The American experience in Vietnam fueled the perception that the vast arsenal of the Cold War was a blunt tool, one that the military could not be trusted to use wisely and humanely. This argument resonated more deeply on college campuses than in the country at large.

Some scientists and engineers, far from renouncing the MIC, warmly embraced it. Charles Stark Draper, for example, actively defended to his MIT colleagues and students the military work of his Instrumentation Laboratory. Edward Teller was perhaps the most strident and influential of these cold warriors. A co-developer of the US hydrogen bomb, he played a notorious role in the revocation of Robert Oppenheimer's security clearance. Throughout the 1960s and 1970s, he condemned the nuclear test ban treaty as unverifiable and the Soviets as mendacious. In the 1980s, he helped convince President Reagan to launch the Strategic Defense Initiative, a program to develop ballistic missile defense.[11] He was reportedly one of the conceptual models that Stanley Kubrick and Terry Southern had in mind for the title role in their 1964 satirical movie *Dr. Strangelove or: How I Learned to Stop Worrying and Love the Bomb*.[12] When American scientist Harrison Brown was negotiating with his Soviet counterparts

in 1960 to organize a conference on the nuclear arms race, a Soviet spokesman said to him, "If you bring your Teller, we will
bring ours."[13] Eisenhower had Teller in mind when he warned of
a "scientific-technological elite."[14]

In general, scientists and engineers lent themselves readily
to the Military-Industrial Complex. While some eschewed military research, many others embraced it as patriotic and stimulating. Most, no doubt, simply looked upon it as a source of
patronage. Researchers are always engaged in an evolving negotiation with their patrons, seeking a balance between their own
muses and the interests of those who pay the bills. In this sense,
patrons always influence the research agendas of the societies
they inhabit. For better or worse, the Department of Defense was
the largest single patron of US science and technology in the
Cold War.

Chapter Six

Society and Technology

The Cold War imposed a unique and perverse logic, not just on strategists, but on the public at large.[1] The superpowers and the other countries of the northern hemisphere resided in the path of the radioactive fallout that would circle the Earth following a "nuclear exchange." They perched precariously on what defense intellectual Albert Wohlstetter called a "delicate balance of terror."[2] Security was achieved through reciprocal vulnerability. This prospect of mutual assured destruction (MAD) chastened both sides to avoid armed confrontation at all costs. When such a showdown arose in the Cuban Missile Crisis of 1962, the experience so alarmed both superpowers that they collaborated thereafter on detente. Each came to realize that its safety was hostage to its enemy's safety.

In such an atmosphere, the normal rules of diplomacy and war were stood on their head. The secrecy surrounding weapon systems was oppressive, yet both sides regularly leaked information to the other. It was essential, after all, that the other side knew and appreciated its enemy's capabilities, the range of its bombers, the accuracy of its missiles, the stealth of its submarines. Promoters of peace advocated the elimination of nuclear weapons, even though the peace itself depended on those weapons. Both sides welcomed the intrusion of reconnaissance satellites, which gave each the assurance that the other was not mobilizing its forces. For many years the United States denied that it used reconnaissance satellites, but the Soviets knew.

The convoluted logic of this dance of death became famously public in the controversy over intercontinental ballistic missile basing in the 1970s and 1980s. The improvement of US missile accuracy had prompted the Soviets to increase both the numbers and accuracy of their missiles, raising in Americans the same fear of an enemy first strike that gripped their adversaries. One way to protect the American MX missile from preemptive strike was to put each missile on a mobile carrier that would move over a closed course visiting 23 different shelters on each circuit. The Soviets would never know which shelter housed a missile or if the missile was still on the carrier. But the placement of the missiles could not be allowed to violate the SALT (Strategic Arms Limitation Talks) treaties of the 1970s, which had limited the allowable number of readily observable weapons. So, the United States would load the carrier with the top open in full view of Soviet reconnaissance satellites. Furthermore, each shelter had viewing ports which would be opened to Soviet observation periodically, or on challenge, to ensure that no more than one missile moved around each course. In a kind of strategic shell game, one side would lift the shell from time to time to let the other count the peas.[3]

Nowhere was the perversity of this relationship more compelling than in the race to build more and better strategic weapons. Perception was everything. It mattered not what a strategic arsenal could really do, only what the enemy thought it could do. Each side had to ensure that its enemy feared overkill, feared that its opponent's arsenal was powerful enough to destroy it several times over. Such an environment built irresistible pressure to err on the side of destruction. No calculus could allow the enemy to believe that he could launch a first strike and survive retaliation from the residue of his enemy's forces; so, both sides built ever more warheads, ever more delivery systems. The largest missiles mounted multiple warheads, ensuring that some bombs from each carrier would reach their target. In time, the multiple warheads were made to navigate independently to their respective targets. They could even maneuver in the face of enemy defenses, though such defenses posed little threat to their flight. At the peak, the two sides brandished more than 10,000 strategic warheads each, even though Defense Secretary Robert McNamara had concluded in the 1960s that 1,000 were enough to destroy the Soviet Union as a viable political and social entity. The technology itself seemed to determine policy. Not for nothing was the strategy called MAD.

The strategic arms race reached a climax in President Reagan's Strategic Defense Initiative (SDI), quickly dubbed "Star Wars" after the popular science-fiction movie. Convinced by a small coterie of advisors that technology could protect the United States from the MADness into which technology had cast the world, President Reagan proposed a layered defense system. It would dome the US with an impenetrable barrier against intercontinental ballistic missiles. Most informed observers in the US and the Soviet Union realized that such a defense was then beyond the current and foreseeable technical capability of even the

US. Again, however, perception trumped reality in the Alice-in-Wonderland world of Cold War logic. Many Americans embraced ballistic missile defense and kept the program going into the twenty-first century in spite of its failure to demonstrate a workable system. Not even the end of the Cold War arrested this development.

Realizing that perception was everything, some opponents of SDI noted that it was fundamentally destabilizing. If successful, it would diminish or eliminate the threat of the Soviet arsenal, tempting the Soviets to "use 'em or lose 'em." Advocates of SDI countered that by raising the ante in the research-and-development contest of the Cold War, SDI would force the Soviets to recognize the bankruptcy of their command economy and begin the transition that led to the collapse of the Soviet Union. Both arguments have merit; both reveal the centrality of technology in this complex calculus and the preeminence of perception. What observers thought about the technology was more important than the technology itself.

In such an atmosphere, it was easy to imagine national leaders who had taken leave of their senses, one of the premises of director Stanley Kubrick's *Dr. Strangelove*. In the memorable, climactic scene, Slim Pickens rides a huge, phallic nuclear missile to an orgasmic suicide symbolic of humanity's insane infatuation with these engines of destruction. The movie's General Jack D. Ripper evoked General Curtis LeMay, the champion of strategic bombing, who favored preventative war with the Soviet Union. Equally plausible was the suspicion that technology itself was in control. In the novel *Failsafe* and its cinematic sequel (1964), a US bomber receives a false and irretrievable command to attack Moscow with its nuclear weapons.[4] Not even Soviet-American cooperation can reverse the doomsday machine that has been set in motion. The technology had escaped human control.

The Cold War and its MAD arms race also contributed to the growth of an alarmist literature about technological determinism. Jacques Ellul claimed in *The Technological Society* that human agency was being lost to machines, which operated by a logic of their own.[5] In the United States, Lewis Mumford echoed this view in *The Myth of the Machine*, a two-volume assault on the Military-Industrial Complex as a triumph of machine over human values.[6] In *Autonomous Technology: Technics Out of Control as a Theme in Political Thought*, Langdon Winner explored the roots of this notion in Western civilization, from Mary Shelley's *Frankenstein* (1818) to Kurt Vonnegut's *Player Piano* (1952).[7] As with most critics of technological determinism, these authors advocated a return of human agency, a conscious campaign to take control of our technology before it destroyed us.[8] Their work, therefore, blended with one of the main themes of postmodernism and poststructuralism, the belief that Enlightenment rationalism had launched Western civilization on the path to an alarming modernity.

Other scholars sounded warnings more closely focused on the MIC. Paul Forman and others lamented the opportunity costs of the Cold War.[9] The billions of dollars invested in weapons research might have been better spent on medicine, social needs, agriculture, and other life-enhancing purposes.[10] This proposition begged the question of whether or not defense funds were fungible, that is, could they have been diverted to humane undertakings. To believe that was to believe that the super powers would have taxed themselves at the same rate and invested the funds more wisely had there been no Cold War.[11] Similarly, Paul Edwards argued that the computer, one of the core technologies behind the rapid transformation of information and communications, was shaped irreversibly by the military sponsorship of its early development. Because the military wanted computers and networks for command and control, they created a "closed world" in which the technology had those goals and values imbedded in its most fundamental structure.[12] The chaotic riot of the internet bore little resemblance to the hierarchical infrastructure of military C^3I (command, control, communication, and information), but the alarm engendered by the excesses of Cold War technology nevertheless provided fertile ground for such conjectures.[13]

In the end, Americans learned to live with the strange paradox that the MIC saved and threatened civilization at the same time. Its strategic arsenal preserved the peace while imperiling all life in the northern hemisphere. Its conventional arms proved brutally effective in the Iraqi desert during the Gulf War of 1990–1991, even while they kept US servicemen out of harm's way in Kosovo. United States military technology failed in Vietnam, in part because the enemy used terrain, tactics, and techniques against which its force was attenuated. However, out of the frustration of Vietnam came the smart weapons that set the world standard in the late twentieth century.

Americans also learned to live with fear. In the 1950s and early 1960s, when nuclear war seemed inevitable, children practiced hiding under their school desks in the event of nuclear attack. Prosperous citizens built bomb shelters in the curious faith that they would want to live in the wasteland left behind by nuclear war. The terrifying devastation of Hiroshima and Nagasaki was compounded by the unfolding nightmare of radiation effects on survivors. Nuclear technology was at once America's greatest technological achievement and its worst peril. Fear of nuclear weapons was projected onto commercial nuclear power, burdening utilities with litigation, safety, and decommissioning costs that far outweighed its economic benefits. The same fear contributed to the early retirement of the nuclear-powered commercial ship N.S. *Savannah*. In time, when nuclear war did not come, and when nuclear power plants did not melt down, the fear subsided. But each time a new state joined the nuclear club, as China did, for example, in 1964, and each time a nuclear power plant experienced a serious "excursion," as Three Mile Island did in the United States in 1979 and Chernobyl did in the Soviet Union in 1986, the old fear revived. In no other technology was the MIC's power to simultaneously protect and threaten so vividly and poignantly manifest.

While living with this fear, Americans also availed themselves of the cornucopia of technology spun off by the MIC. The aerospace industry rode defense dollars to world preeminence. Consumer electronics exploited the solid-state devices first developed for weapons and space applications. Major corporations such as General Electric, Motorola, and IBM kept one foot in the military marketplace and one in the civilian, finding cross-fertilizations that advanced technology in both. The Jeep, workhorse of World War II personal transportation, found itself transformed into a hot commercial product for upscale adventurers. Its distant successor, the army's HMMWV (High Mobility Multipurpose Wheeled Vehicle) inspired the commercial "Hummer," a status symbol for those with $70,000 or more to invest in an automobile wider than many parking spaces. By the end of the twentieth century, the military's global positioning system (GPS) was guiding everything from commercial ships and airliners to private yachtsmen and hunters and even drivers of Cadillacs equipped with the company's "NorthStar" system. Computer speech recognition programs developed for the military found their way into interactive weather reports by telephone, airline reservation systems, and even hotel guides to restaurants and other local businesses. For better or for worse, the products of the Military-Industrial Complex found their way into almost every corner of modern life, to be welcomed openly, if often unknowingly, even by those who once took President Eisenhower's warning as prophetic.

Chapter Seven

International Arms Trade

T he international arms trade existed long before the Military-Industrial Complex, indeed, long before the United States. In both world wars, the United States acted as an "arsenal of democracy" and profited from the service. At the same time, after World War I, it introduced the epithet "merchants of death" to stigmatize Americans who profited from the war. In the run-up to World War II, it cloaked its own profiteering in the euphemism "lend-lease," a term that masked the high price exacted from Britain for US arms and equipment before the US entered the war. The same locution calmed the protestations of "America-firsters," who wanted the US to stay out of World War II. So Americans entered the Cold War arms bazaar with their eyes wide open. The euphemism of choice in this context was "arms transfer."[1]

Arms transfers were primarily instruments of Cold War competition with the Soviet Union and its allies in the Warsaw Pact. The transfers attracted allies to the United States by providing military arms and equipment that were arguably superior to comparable products from the communist bloc, and they bound these customers to US manufacturers for parts and servicing.

They facilitated military cooperation between allies using similar or even identical equipment. They seduced neutral states in the Cold War competition for the hearts and minds of the so-called "Third World." Further, they subsidized the US arms race with the Soviet Union by recovering costs of R&D and production. The more products sold, the lower the unit cost to the US for its purchases from industry. And, of course, the producers thrived on additional sales after recovering their own investments in capital plant and R&D.

At the same time, there were reasons to limit international arms sales. The spread of conventional arms promoted war and carnage around the world. In spite of constraints imposed by the United States, purchasers often used US weapons and equipment for purposes condemned by America. Once the weapons were delivered, the US often lost control over their use against civilians, innocents, and even other recipients of US arms. However, in the end, reasons of state usually trumped moral and ethical qualms.

The market quickly became oligopolistic, with the United States and the Soviet Union producing most of the arms and making up to 80% of the transfers. The purveyors also restricted the technology that could be exported in such transfers, even as they asserted more control over non-weapons technologies, such as computers, machine tools, microelectronics, communications equipment, space technology, and the like. Strictly monetary trades gave way increasingly to transfers, in which the purchasing state could negotiate licensing agreements, coproduction schemes, technology-sharing, and other "offsets" that might advance their own technological development. All such effects obfuscated the money value of the transfers and made it more difficult to measure the volume of the international arms trade. Political constraints on transfers bred growing "black" and "gray" markets in which transfers escaped—or partially escaped—public disclosure. In such markets, criminals and other non-state actors began to acquire significant arsenals on a scale previously confined to states.[2]

Of course, the greatest danger after World War II arose with the prospect of nuclear weapons. In the 1940s, four states achieved nuclear capability. Britain and France acquired nuclear weapons from their collaboration with the United States in the Manhattan Project. Soon, the Soviet Union joined them, aided in part by espionage. China became the fifth nuclear power in 1964, with help from the Soviet Union. In the late 1960s, those five states—the permanent members of the United Nations Security Council—collaborated in drafting and passing the "Treaty on the Non-Proliferation of Nuclear Weapons," which went into effect in 1970. By that time, Israel had already—though secretly—joined the "nuclear club," with help from France, Great Britain, the United States, and Argentina. India joined in 1974, spurring its neighbor and archenemy Pakistan to accelerate its own bomb development.[3] From that campaign emerged the commercialization of nuclear weapons capabilities, though not before the end of the Cold War.[4]

Trade in conventional arms proceeded quite differently.[5] In one of the many ironies of the Cold War, the flow of conventional arms around the world accelerated even as deaths from war declined steadily, even precipitously.[6] The superpowers and their allies built up enormous arsenals to confront each other across the globe, with special attention paid to the prospect of a land war in Eastern Europe. The Soviet Union and its allies relied on the quantity of its troops and arms. The United States and its allies banked on superior weapons and equipment. The resulting arms race witnessed the diffusion of Soviet and US weaponry through the Warsaw Pact and NATO and joint weapons development pacts between the super powers and their allies. Secondary states within the armed camps of the Cold War—Czechoslovakia, for example, and Great Britain—developed production capabilities that weighed in the world market, while nonaligned states such as Sweden and Switzerland produced arms for sale to either camp.

The international market for conventional arms mixed sales and transfers in a complex web of economic and security motives. Neutral states sold their wares almost entirely for economic reasons, even though they contributed to a warring world they often claimed to deplore. The states aligned in the Cold War responded to economic motivations as well. Sale of the obsolescent weapons that they had paid to develop helped amortize their military costs and underwrite the next generation of arms and equipment. Critics in the United States even claimed that the US economy was dependent on the arms industry.[7] Other observers, equally critical of the Military-Industrial Complex, argued that military expenditures actually retarded economic development by investing capital and labor in products that did not add value to the civilian economy.[8] Still, the imperatives of the arms race drove the industrialized states to continuing improvements in armaments and a resulting surplus of obsolescent materials. The temptation to sell these abroad proved irresistible.

Complementing the sale of these weapons for economic gain,

many industrialized states, especially the two superpowers, also transferred their weapons abroad in deals motivated by strategy and politics. The Truman doctrine, for example, promised military aid to "free" states resisting foreign—that is, communist—intervention or provocation. Arms transferred to states under this rubric changed hands under all sorts of conditions. Sometimes they were given outright, though always with political strings attached. Sometimes they were sold under loans or discounts. Almost always they bound the receiving state to the patron by the need for replacements, repairs, and spare parts. Sometimes one or both of the superpowers almost completely armed the combatants in proxy wars such as Korea (1950–1953), Vietnam (1945–1975), or the Arab-Israeli conflict, especially the Yom Kippur War of 1973. From time to time, the superpowers even found their weapons in the hands of combatants on both sides of the same conflict, as, for example, when the Indians and Pakistanis both used US and Soviet weapons in their wars of the 1970s and 1980s. Always the states involved in these arms transfers exchanged weapons for something of value—money, political alignment, proxy conflict, territorial or sovereign concessions, or bondage to the supplier—but seldom did market forces dictate the nature of the transactions.[9] Commodities changed hands, but the currency of this realm defies classification. For that reason, it is difficult to quantify the volume or dollar value of the trade.

Offsets compounded the complexity of arms transfers. They provided the receiving state a kind of kickback on the nominal price of the purchase.[10] Offsets included licensing agreements, joint manufacturing schemes, and training for the maintenance and repair of equipment. And, of course, time-honored traditions of bribery lurked in the shadows of every negotiation.[11] All such offsets offered buyers some economic or technological advantage. Offsets became important tools in the Cold War as the superpowers competed in their arms sales for the allegiance of so-called third world countries. The Cold War competition, paired with the proliferation of arms producers, made it a buyer's market.

Meanwhile, purveyors of arms attempted to limit and direct the flow of goods and technologies to hostile states. The United States, for example, authorized transfer of its obsolescent technologies, such as fighter aircraft and Stinger anti-aircraft missiles, to friendly states, but resisted pressure to share its state-of-the-art equipment with any but its NATO allies. It also insisted that states receiving its equipment and technology not share them with third parties, a requirement that proved hard to enforce. Indeed, by the end of the Cold War even Western nuclear weapons technology was seeping out of the nonproliferation regime and spreading among rogue states such as Pakistan, Libya, Iran, and North Korea.[12] The so-called "black market" in arms flourished, frustrating attempts to track and calculate the volume of arms sloshing about the world marketplace. Furthermore, the "gray market" in arms flirted with legal restrictions, most infamously in the US "Iran-Contra" scandal. US government officials secretly sold weapons to Iran through an intermediary— Israel!—in exchange for funds to support anti-communist rebels in El Salvador, a transfer of military aid specifically banned by the US Congress.[13]

All kinds of arms changed hands. From 1967 to 1976, major suppliers provided to developing countries alone 6,212 combat aircraft; 3,026 noncombat aircraft; 2,972 helicopters; 113 major surface combatant vessels; 640 minor surface combatants; 54 submarines; 18,607 tanks and self-propelled guns; and 19,376 armored personnel carriers and armored cars.[14] Still more transfers of even more sophisticated equipment moved between the superpowers and their allies in the great alliances of NATO and the Warsaw pact.

The politics of arms transfers, and the numerous evasions concocted to restrict public disclosure, made accounting for the international arms trade almost impossible.[15] It can be said with some confidence that the United States and the Soviet Union accounted for most of the transfers—something on the order of 65–80%—through most of the Cold War, with the United States being the leading exporter until the late 1970s and the Soviet Union dominating thereafter until its collapse around 1991. A group of "second-tier" arms manufacturers—led by France, Britain, West Germany, Czechoslovakia, and later China—accounted for 20–30% of transfers. The four-year average dollar value of the global trade (in 2019 dollars) ranged from $27.4 billion (1963–1966) to $117 billion (1983–1986).[16] These totals accounted for about 2% and 1.4% respectively of global world exports.[17]

Since the Cold War, 1991–2020

Chapter Eight

New World Order

No Peer Rival

Toward the end of the twentieth century, the leaders of both the United States and the Soviet Union began speaking of a "new world order." Soviet President Mikhail Gorbachev invoked the phrase in an address to the United Nations in 1988, as the Soviet Union was entering its death spiral. President George H. W. Bush embraced it in a speech to a joint session of Congress on September 11, 1990, on the eve of the first Gulf War. By the end of 1991, the Iraqi army had been driven from Kuwait all the way to Baghdad and the Soviet Union was no more. Students of world history and American national security policy were left to debate whether the "new world order" was an American blueprint for the post–Cold War world or simply an evocative slogan to mark a sea change between historic epics.[1]

Surely the collapse of the Soviet Union had the greater impact on the Military-Industrial Complex. The MIC had arisen from the standing military establishment raised to meet the challenge of Soviet ideology and expansionism. It had taken on its characteristics and practices in the nuclear and conventional arms races between the two superpowers. In the minds of many, that contest had driven the United States into the trap warned against by George Kennan. He had cautioned that perceiving the Cold War as an existential conflict could induce the United States to remake itself in the form of its enemy. This possibility echoed Harold Lasswell's apprehension that the US might, in the grip of fear, become a garrison state, a concern that President Eisenhower came to share.[2] The Reagan administration rekindled such fears in its first term, raising defense spending and reviving apprehensions of Armageddon. The Military-Industrial Complex grew in that final spasm of Cold War escalation and then declined for the rest of the twentieth century.

The Gulf War of 1990–1991 also played an important role in America's path into the post–Cold War world. The conventional arsenal built up by the United States to confront the Soviet Union on the plains of Europe also seemed well suited to the Middle East, where the United States had found itself increasingly entangled. Americans consistently saw national interests in the region that they could not ignore or escape, from the CIA-assisted overthrow of the democratically elected prime minister of Iran in 1953 and the Suez Crisis of 1956 to the Arab-Israeli wars of 1967 and 1973, the Iranian Revolution in 1979, the continuing courtship of Saudi Arabia and its oil, the OPEC embargo of 1973, and the Iran-Iraq War of 1980–1988. The United States repeatedly deployed its military, diplomatic, and economic power, always with an eye to checking Soviet strategic ambitions. Sooner than anyone expected, the ongoing saga of American involvement in the Middle East would shape the demands it made upon its Military-Industrial Complex.

President Reagan's successor, George H. W. Bush, presided over America's triumph in the Cold War and the Gulf War of 1990–1991, strong credentials for reelection in 1992. In that contest against Democrat Bill Clinton, two major issues dominated public discussions of national security. First, and most salient with the voting public, was the "peace dividend." Americans had born the burden of a large standing military establishment in peacetime for more than four decades. The Cold War was over, and they had won. Why not demobilize, as America had done after its previous wars, and return to the constabulary force necessary to protect the homeland and America's vital interests abroad? The answer, of course, was that the United States had designed the reigning liberal world order, institutionalized in the United Nations and its partner organizations such as the World Bank and the International Monetary Fund. It had also allowed itself to become the world's policeman, a role dictated by its own hegemonic ambitions. The isolationism it had embraced after World War I had allowed—even invited—the totalitarianism of the Axis powers in World War II. To withdraw again behind Fortress America risked a similar resurgence of forces inimical to American interests. To sustain the Pax Americana that began in 1945, it would be necessary to maintain some standing military establishment and some kind of defense industrial base (DIB) to arm and equip it. But without the existential threat posed by the Soviet Union, it might be hoped that a smaller establishment and a more modest DIB could serve.

The second security question facing the country in 1992 was what grand strategy the United States should embrace to nurture the new world order. Might the world be ready to abolish nuclear weapons altogether, as both Presidents Gorbachev and Reagan had counseled? The American public and many world leaders said yes; the American military and many security experts said no. The "no's" won, preferring the proven effectiveness of deterrence to the uncertainty of conventional military competition. Still, the nuclear arsenals of the Cold War could be further reduced, and the nonproliferation regime could be strengthened. As for conventional weapons, the military community agreed that the world was still a dangerous place, populated by many unfriendly states and institutions. The United States still needed an armed force equal to those many threats. The guarantor of American military preeminence remained, in the minds of both military and civilian leaders, an arsenal of unmatched technological superiority.

The answers to these two imperatives—a peace dividend and a grand strategy—would determine the fate of the Military-Industrial Complex. No one doubted that it had existed during the Cold War, that it had demonstrated many of the characteristics and problems identified by President Eisenhower in his farewell address, and that it had played a significant role in America's victory. It had certainly produced the necessary arms and equipment, even while it inflicted upon the United States the penalties, frauds, burdens, and excesses that Eisenhower had warned about. In short, it had served America well in an existential crisis, while exacting a high price. Should the country continue paying that high price? In the absence of an existential threat, had the raison d'être of the MIC also passed away?

The transitional presidency of George H. W. Bush offered conflicting answers. The collapse of the Soviet Union at the end

1991 was unexpected and precipitous. The disintegrating super power fell into political turmoil between 1989 and 1991; defense spending fell by 73% in one year (1991–1992) and a total of 89% between 1987 and 1995.[3] The member states of the former Soviet Union and Warsaw Pact took their weapons, their allegiances, and their futures in multiple different directions. If the Soviet Union had ever posed an existential threat to the United States, it all but disappeared overnight, save for its aging and diminished—though still potent—arsenal of strategic nuclear weapons. At the same time, however, the United States found itself drawn into the Gulf War of 1990–1991, leading a coalition of mostly NATO states in a military campaign to force Saddam Hussein to withdraw from his invasion of neighboring Kuwait. Unless the United States wanted to retire from its position as leader of the free world and perhaps guarantor of the new world order, it would have to maintain some semblance of the military establishment and the DIB it had mobilized to win the Cold War.

The Bush administration's answer to this pivotal question appeared in a classified, preliminary, internal document called the "Defense Planning Guidance" of 1992. It was known unofficially as the Wolfowitz doctrine, after Paul Wolfowitz, the under secretary of defense for policy.[4] Its first priority was "to prevent the re-emergence of a new rival. . . . We must," it said, "maintain the mechanism for deterring potential competitors from even aspiring to a larger regional or global role."[5]

When this draft leaked to the press, the *New York Times* interpreted it as prescribing "benevolent domination by one power" or "Maintaining a One-Superpower World."[6] A public outcry forced the Bush administration to tone down the rhetoric. In private, however, defense hawks continued to embrace the principle of "no peer rival" as the most fundamental tenet of post–Cold War American grand strategy.[7] The United States would remain the world's sole superpower in a multipolar world. It would retain sufficient nuclear and conventional weapons to deter and, if necessary defeat, any other state or coalition of states that might threaten its national security. Of course, it would continue to work with its Cold War allies—essentially the Western democracies plus Japan, South Korea, and other states making common cause with this band of collective security—but it would remain the first among equals. And it would maintain a military establishment and a DIB sufficient to guarantee that position.

The "Defense Planning Guidance" of 1992 was offensive to many Americans because it seemed to perpetuate the alarmist and aggressive worldview that drove up Cold War military spending, and it seemed to have been concocted in secret and embraced by a coterie of hawkish policymakers who could well be in league with the MIC.[8] Indeed, one of the hallmarks of the MIC had been the suspicion that its members secretly conspired to push the country into expensive, dangerous, and unnecessary military expenditures and adventures. This was just the sort of paranoid and conspiratorial Cold War thinking that many Americans wanted to exclude from the new world order.

So the explicit policy of "no peer rival" disappeared from official, public discourse; but it remained an unspoken aspiration of the US military establishment, the DIB that served it, and congressional hawks advocating increased defense spending. Never again did people in those communities want the United States to face an existential threat, as they believed it had during the Cold War. Military and technological superiority had brought the country through that challenge, and the same formula could work in the future. Or so they believed.

What they failed to appreciate was that their share of the world's economy was shrinking. The United States still had the world's largest and strongest economy—commanding more than 26% of world GDP in 1992 with only about 5% of the world's population—but that figure was down from 56% of world GDP in 1950, when the Cold War began.[9] Might some state or alliance one day match America's buying power and build its own defense industrial base of equal size and quality? The combined military assets of the US and the countries aligned with it seemed proof against that danger, but members of the national security establishment were inclined to count threats, not allies. The next contender for peer rival might have an economy to match America's.

The challenge for the Bush administration was to politically balance its embrace of "no peer rival" with the domestic, political imperative to provide the American taxpayer with a peace dividend. The collapse of the Soviet Union and the deterioration of its arsenal meant that the United States could probably maintain its global preeminence with lower levels of military spending. The growth of the US economy meant that even higher levels of military spending could be realized with smaller percentages of the GDP and the growing federal budget. Or at least slower growth in military spending could be safely tolerated.

And so defense spending fell steadily from 1989 to 2000.[10] In current dollars—that is, then-year dollars—defense spending increased above the peaks for the Vietnam War, the Reagan buildup, and even the Korean War, which was previously the

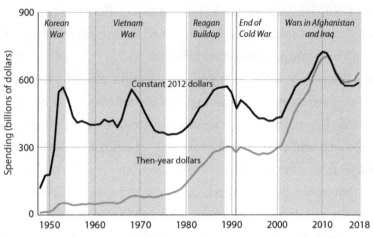

CHART I. US military spending in current and constant (2012) dollars, 1948–2018. *United States, White House, Office of Management and Budget,* Historical Tables, *table 6.1.*

highest level of military spending since World War II (chart 1). But as a percentage of the federal budget, military spending has declined from 37.1% in 1947 to 15.5% in 2018, having reached a high of 69.5% in 1954. Similarly, military spending as a percentage of GDP went from 5.4% in 1947 in to 3.1% in 2018, having reached a high of 13.8% in 1953 (chart 2). These trends allowed the United States to stay far ahead of the rest of the world in annual military spending while actually lessening its defense burden as a share of the total economy and as a share of the federal budget. Military spending rose, but not as fast as the wealth of the nation and the size of its government.

The implications of this pattern for guns and butter were profound.[11] The United States was so wealthy that it could have both guns *and* butter, or swords and plowshares, as Ron Smith prefers to phrase it.[12] Congress and successive administrations

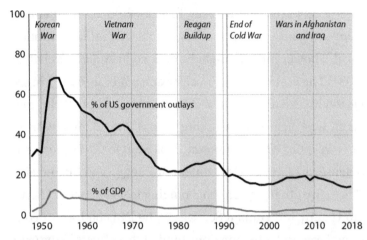

CHART 2. US military spending as a percentage of GDP and federal budget, 1948–2018. *United States, White House, Office of Management and Budget (OMB), Historical Tables, table 6.1.*

repeatedly took the easy way out by simply refusing to choose between them. There was always enough money in the economy to do both. When revenue fell short, deficit spending filled the gap. The government chose not just more guns and more butter, but *lots* more of both.

The Technological Fix

Still, the greatly reduced defense budgets of the 1990s were real—in constant dollars. They represented something like a recession within the defense marketplace, calling for drastic changes in all corners of the Military-Industrial Complex. Lacking a great-power rivalry to compare to the Cold War confrontation with the Soviet Union, the United States nonetheless resolved to maintain a standing military establishment capable of deterring any strategic nuclear threat while simultaneously meeting any conventional challenge to its vital interests around the world, however those might be defined by America's civilian leaders.[13] The cost of the US nuclear arsenal is shrouded in secrecy, but an exhaustive study in 2009 concluded that the government was then spending more than $52.4 billion a year on nuclear weapons. Of that amount, 65% went to the Department of Defense, 30% to the Department of Energy, and the remaining 5% to five other departments ranging from Homeland Security to Health and Human Services.[14] The DoD budget for conventional war and other missions funded three "wars"—the Gulf War (1990–1991), Iraq War (2003–2011), and the war in Afghanistan (2001–?), plus the so-called "War on Terror," (2001–?). Additionally, the US military carried out perhaps nine overseas "conflicts" in which American servicemen and women died, in addition to countless humanitarian and peacekeeping missions and police actions, such as guarding commercial vessels in the Strait of Hormuz or "Freedom of Navigation" passages through waters claimed by China in the South China Sea.

Even deficit spending, should Congress continue to allow it, would not make up for the peace dividend. Somehow the DoD would have to do more with less. As if on cue, a "Revolution in Military Affairs" appeared to offer a solution. Its advocates claimed that it would leverage American technological advances to project military power more efficiently, more effectively, and more safely than any other country. They offered, in short, a technological fix to the peace dividend.

The roots of this solution ran deep into the Cold War. The Carter administration had faced a similar dilemma in the late

1970s. The American public also demanded a peace dividend after the costly and divisive Vietnam War, even though the security climate was darkened by a Soviet military buildup. To make matters worse, the US economy was strained by high inflation and slow growth. Carter's Secretary of Defense Harold Brown, a scientist, proposed a second offset, a policy consciously modeled on what Eisenhower had called the "great equation": maintain military superiority without ruining the American economy.[15] The United States would use its scientific and technological preeminence to "offset" Soviet superiority in numbers. In short, they planned to revitalize the nation's commitment to quality over quantity, embracing the technological fix to shore up the Pentagon's bruised reputation after Vietnam while living within severe budget constraints.[16] Brown teamed with Under Secretary William Perry, a mathematician, to execute the policy.

Brown and Perry looked to precision-guided munitions to correct the deficiencies of bombing and close-air support in Vietnam. They encouraged the army's emerging concept of AirLand Battle to exploit American strengths in communications, computers, sensors, and other electronics. And they provided an intellectual environment in which retired air force colonel John Boyd could preach his OODA theory of air combat: observe, orient, decide, act. High-speed gathering and processing of information, said Boyd, allowed quick decisions and preemptive action. Technology could enable US pilots—and all combatants—to get inside the enemy's "decision-making cycle"—his "OODA loop"—and shoot first.[17] Visions of an "electronic battlefield" of the future portrayed Americans ranging at will and projecting force with impunity.[18]

The "second offset" stimulated a wave of research and development, especially in microelectronics, and it complemented a parallel military reform movement in Congress and the private sector. The Congressional Military Reform Caucus, formed in 1981, focused on problems in military acquisition, seeking efficiencies in development and procurement.[19] But, like the Brown/Perry offset, its impact was swamped in a wave of defense spending in President Reagan's first term (1981–1985) (figure 4). Budget increases in those years did include significant growth in research and development, but the simultaneous expansion of personnel, acquisition, and operations budgets lessened the urgency of the "offset" and reform.

Still, by the middle of the 1980s, a clear trend in American military technology was taking shape. The American paradigm of military superiority in World War II—industrial productivity and technological innovation packaged in massive weapon systems—was giving way to a "defining technology" of warfare after the Cold War. As historian J. David Bolter argued in his history of the emerging computer age, epochs in history can organize themselves around a defining technology that captures the essence of humans' efforts to bend nature to their will.[20] If, as Walter Millis argued, the internal combustion engine was the defining technology of World War II, and if nuclear weapons were the defining technology of the Cold War, then the computer or information processing or networking was becoming the defining technology of the post–Cold War.[21] With greater perspective from the twenty-first century, it might be more accurate to say that the transistor—avatar of microelectronics—is a better icon for the phenomenon, but the idea is the same.[22]

FIGURE 4. Two early transistors (one uncovered), huge by the standards of their microminiaturized descendants, reveal their size by comparison with the nearby US dime. *Mark Wichary, Flickr, licensed under CC BY 2.0, https://creative commons.org/licenses/by/2.0/.*

Several authors have wrestled with alternative characterizations of this defining technology. The most important innovations are especially concentrated in what might be called a "microelectronics domain." Ann Markusen and Joel Yudkin called it ACE (aerospace-computer-electronics). They credited Herman O. Stekler with first discerning the impact of the aerospace industry in promoting these technologies.[23] Linda Weiss, who wrote of a "national security innovation engine," cites Vernon Ruttan's listing of aerospace, computers, and semiconductors as hallmarks of "America's leading high-technology commercial sectors."[24] Rebecca Thorpe has discerned an MIC "strategic shift to procurement of radar equipment, telecommunications, and electronics" in the 1990s.[25] This sea change in the focus of military innovation may also be seen as a shift from the macro weapon systems of the Cold War to the micro world of the electron. Aerospace technologies led the way because they were so dependent on weight-saving, force-multiplying microminiaturization.

These microelectronic technologies based on semiconductor devices and integrated circuits have transformed the modern quest for superior military arms and equipment. The technologies that would allow John Boyd to get inside the decision-making cycle of the enemy were sensors, computers, artificial intelligence, and servo-mechanisms that were relentlessly becoming smaller, lighter, faster, and smarter. And of special appeal to Americans, such gadgets could cross the battlefield autonomously or remotely controlled while American servicemen and women remained out of harm's ways.

As the salad days and lavish budgets of the Reagan defense buildup waned in his second term, belt-tightening resumed, and cost-savings regained their purchase on the military imagination. The contributions of the second offset and the military reform movement, combined with a growing concentration on research in the "microelectronics domain" coalesced in the Revolution in Military Affairs (RMA).[26]

This embodiment of the post–Cold War defining technology

promised unprecedented and incomparable military capability at bargain prices. Interception of enemy attacks meant fewer casualties and less damage to material assets. Superior intelligence and communication might even allow preemption of such attacks. Greater precision in American weapons meant more damage from less munitions. Microminiaturization promised to be the greatest of all "force multipliers," giving the United States an invincible arsenal at low cost. The MIC already had sufficient inroads into this technological treasure trove to ensure continued leadership. The RMA offered an irresistible argument for military personnel and their civilian masters, because it meant that the United States could buy more military power for less. The country could give the voters a peace dividend while the military actually increased its fighting power.

Not surprisingly, the RMA had as much to do with Pentagon politics as it did with American strategy. With the demise of the Soviet Union, many reformers perceived that the MIC would have to be redirected from the great, expensive weapon systems of the Cold War—aircraft carriers, strategic bombers, ballistic missile submarines, tanks, intercontinental ballistic missiles, armored personnel carriers, even fighter aircraft—to the tools of "low-intensity conflict," or "asymmetric warfare." This meant small wars, interventions, anti-insurgent actions, even peacekeeping and state building, in which the United States would encounter comparatively simple, crude, unrefined weapons, as it had in the jungles of Vietnam. Only now its sensors, communications, and precision-guided munitions would prevail.[27]

But the Military-Industrial Complex of the Cold War was an institution of enormous momentum, a culture within which in-

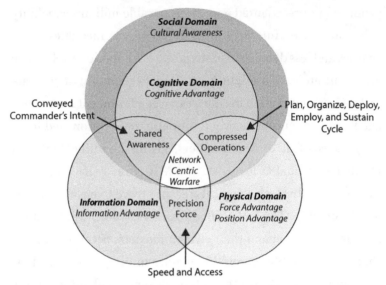

FIGURE 5. This graphic representation of net-centric warfare places it at the center of four "domains"—social, cognitive, informational, and physical. It echoes John Boyd's notion of getting inside the enemy's OODA loop. *Frank Hill, "The Challenge of Simulation Support for Network Centric Exercises," https://www .sisostds.org/DesktopModules/Bring2mind/DMX/API/Entries/Download?Command=Core _Download&EntryId=26168&PortalId=0&TabId=105.*

dividual, corporate, political, and bureaucratic identities had been constructed around those monstrous weapon systems and their attendant infrastructure. All the faithful within the MIC had thrived within that paradigm. Would they or could they adapt to fit a new world order and the transformational arsenal on which it would depend? And was the old arsenal to be replaced with the new "electronic battlefield" or "net-centric warfare" (as it came to be called), or simply complemented with a new tool box designed for a new kind of warfare (figure 5)? And how would the DIB of the new world order compare with the DIB of the Cold War? Would it still function in a Military-Industrial Complex?

Peace Dividend

Consigning "no peer rival" to the realm of the secret and unspoken, the George H. W. Bush administration published its National Security Strategy (1993), a blueprint for securing the nation and policing the world with diminished resources.[28] It recommended a military establishment of 1.6 million servicemen and women, down from 2.2 million in 1989.[29] It spoke of maintaining and expanding existing alliances, while also being prepared to go it alone if necessary. It spoke of proactive interventions to maintain the current "world order" as it saw fit. Implicit in the whole statement was a need to arrest, if not reverse, the cuts to defense spending that had been going on for seven years. It implied that the United States should continue to rely upon the preeminence in arms and equipment that had been the hallmark of its Cold War strategy. And it recommended a military capable of projecting American force to Europe, East Asia, the Middle East, and anywhere else the world order might be threatened. Behind all the plans for ever more refined technology was an emerging commitment to limit "boots on the ground"—that is, the placement of American servicemen and women in harm's way—by substituting high-tech, standoff weapons that could target enemies accurately at a distance. This blueprint of American military policy, to project power while minimizing military casualties, would shape American grand strategy and demands on the MIC for arms and equipment.

The challenge of meeting these commitments with diminished resources had different implications for the different services.[30] The navy lowered its sights from the 600-ship fleet proposed in the first Reagan term to something in the low 300s. At the heart of this fleet would be 10 or more nuclear-powered aircraft carriers—down from a maximum of 15 in 1975—the most salient platform for projection of American power around the world. Complementing these Cold War behemoths of the high seas with their attendant support vessels and submarine escorts would be revolutionary littoral vessels, designed, some with stealth technology, to operate in coastal waters, and high-tech Zumwalt-class destroyers—nimble, all-purpose platforms for an array of sensors and precision-guided munitions.

For the air force, the new world order meant, first, sustaining the world's dominant air-superiority fighter and, second, developing a next-generation strategic bomber to replace the born-again B-1 and the 40-year-old B-52. In addition to the B-2 stealth bomber, replacement was needed for the controversial but world-beating F-22 Raptor—a flawed gem of limited utility—and the pioneering but aging F-117 stealth fighter. The Raptor, coming into service late and over cost just as the Cold War—it's raison d' être—ended, became an expendable "fighter without a foe," as Nick Turse called it, a symbol of the cost and anguish entailed in transitioning from a familiar Cold War to an uncertain future.[31] By 1992, the air force had committed itself to the F-35 Lightning II Joint Strike Fighter, which would become the most expensive weapons system ever developed.

The army, the least technological of the three major armed services, had its own problems giving up the legacy weapon sys-

FIGURE 6. The long cannon barrel of the massive Crusader 155 mm artillery system belies its army classification as a howitzer, normally a stubby, short-range, high trajectory weapon. *David Stubbington, Flickr.*

tems of the Cold War in favor of the microelectronic technologies envisioned by the Revolution in Military Affairs.[32] It had taken a first step in the 1980s with AirLand Battle, a scheme to counter superior Soviet numbers with new technologies beginning to issue from the microelectronics domain. But the army's first major weapon system proposed after the Cold War was the Crusader self-propelled 155 mm artillery system, still designed to meet those Warsaw Pact forces on the plains of Europe, not low-intensity conflicts against irregular, indigenous forces (figure 6). The gold-plated, $11 billion Crusader program experienced development problems from the very start and suffered cost overruns, redesigns, and program delays throughout its seven-year history.

The marine corps, the leanest of the military services and the least likely to succumb to technophilia, found itself plagued by the teething problems of the Osprey vertical and short takeoff and landing aircraft (VTOL/STOL).[33] Developed in response to the tragic collision of a helicopter and a fixed-wing transport in the failed rescue attempt of American hostages in Iran in 1980, the Osprey was supposed to take off and land vertically or on a very short runway, but fly horizontally like a conventional, fixed-wing aircraft. The marine corps already had a long and troubled history with adoption of the British Harrier VTOL aircraft, which experienced one of the highest accident and fatality rates of any military aircraft through the 1960s and 1970s, before technical and operational reforms finally rendered it tolerable. The Osprey, assigned to both the marine corps and the air force, proved similarly accident-prone and dysfunctional in its early years and never achieved the combat effectiveness envisioned for it. It has been used primarily in support roles. Meanwhile, the marine corps bet on yet another VTOL aircraft, a hybrid of the controversial F-35.

When President Bill Clinton denied George H. W. Bush a second term in 1992, it signaled a change in some, but not all, of these priorities and policies. Least surprising was the continued decline in defense spending, bringing into view a rollercoaster pattern to military budgets that would continue through the post–Cold War era. Three Republican presidents (Ronald Reagan, George W. Bush, and Donald Trump) began their presidencies with dramatic increases in defense spending. During the administrations of Democrats Bill Clinton and Barack Obama, defense spending declined sharply.[34] This pattern of partisan policy shifts hampered long-range military planning and complicated the problems already inherent in long-term defense contracting.

Of course, the participants in the Cold War Military-Industrial Complex worked to promote the policies of the Republican presidents and to frustrate the policies of the Democratic presidents. The net result, it appears, was a worsening of the inefficiencies of the system without any compensating advantages. The long suit of American representative democracy is not efficiency.

Military personnel levels fell by 18% between 1993 and 2001, from 1.7 million to 1.4 million.[35] Economies sought in military base closings were favored by the military, inconsequential to industry, and upsetting to politicians.[36] Attempts to cancel contracts for major weapon systems often met resistance from all three branches of the MIC—the military, industry, and Congress. The result was a very democratic muddle in which rather steady reductions in defense spending lurched from one cutback to the next, generally following the trajectory sought by the Clinton administration but making exceptions dictated by Congress, the military, and industry special interests. As historian Harold Parker was fond of saying, out of the conflict of wills emerged a result that no one had willed.[37]

Conversion, Consolidation, Contraction

For its part, the DIB adopted a range of strategies to cope with the shrinking budget of the 1990s. Most of them entailed some combination of conversion, consolidation, or contraction. General Dynamics, the number two ranked defense contractor at the end of the Cold War, at first embraced contraction. Reliant on the military for 90% of its sales and holding a portfolio that included tanks, missiles, launch vehicles, the F-5 and F-16 fighters, and nuclear-powered submarines, the company found itself perilously overcommitted at the end of the Cold War. Choosing to shore up its financial position and protect its investors and senior executives, it sold off assets to companies better able to weather the drought. Its F-16 program went to Lockheed, its missiles to Hughes Aircraft, and its space launch business to Martin Marietta.[38]

It retained its tanks and submarines. It was the only American manufacturer of tanks and it shared all submarine manufacture with just one other company, Newport News Shipbuilding, the sole builder of American aircraft carriers. General Dynamics, originally a spin-off of Electric Boat, the submarine builder, was created to develop a diversified portfolio and act as an integrator of DIB capabilities. Integrators were diversified corporations that sought leverage in large and complementary portfolios, wide experience in government contracting, and sometimes vertical integration within all or part of a weapon system.[39] In the hard times of the 1990s, General Dynamics trimmed its sails to concentrate on its core competencies and its largest project—29 Seawolf nuclear attack submarines (figure 7).[40] The plan worked for General Dynamics, not so much for the navy and the federal budget.

Work had been scheduled to begin on the first Seawolf in 1989, with two more to be laid down in 1991, just as the Cold War was ending. In 1990, the Bush administration proposed cancelling all Seawolfs except for the one already started. The expensive Seawolfs, designed mostly to track Soviet submarines, were no longer needed. Furthermore, the program had already experienced significant cost overruns, due to gold-plating, prema-

FIGURE 7. USS *Seawolf*, first of its truncated, eponymous class of nuclear-powered fast attack submarines, conducts sea trials on September 16, 1996. The end of the Cold War curtailed production of these boats, precipitating a political struggle over contract cancellations and job losses. *National Museum of the US Navy, prepared by Jim Brennan of the Office of Public Access, National Archives and Records Service II.*

ture construction, and repeated change orders. The original 1980s cost estimate of $1.3 billion each for 29 submarines swelled to $5.3 billion in the end.[41] With the navy's approval, all were to be replaced in the late 1990s by new Virginia-class submarines— smaller, cheaper, stealthier, and equipped with better sensing and detection technology.[42] The cancellation of most Seawolfs posed an existential threat to Electric Boat and General Dynamics. Like many other major contractors in the MIC, General Dynamics had expanded optimistically in the salad days of the early Reagan administration, from 1981 to 1986, adding plant and distributing profits at an improvident rate. When the downturn in defense spending began in the mid-1980s, General Dynamics was overextended and further handicapped by years of injudicious management. Having just narrowed its portfolio, General Dynamics could ill afford to hang on until the beginning of the Virginia-class program. To make matters worse, "the company's name [had become] synonymous with cost overruns, fraud, and bill-adding," according to Rachel Weber. [43]

Nothing daunted, the MIC went to work, deploying familiar arguments from the Cold War. Thousands of jobs were at stake, not just at Electric Boat in Groton, Connecticut, but at subcontractors around the country. Senators and congressmen from the regions hosting those subcontractors were warned of lost income and jobs in their districts and states. Other legislators were warned that their defense industries could be next. All were reminded that the new Russian Federation still retained an enormous nuclear arsenal and a fleet of nuclear submarines second only to that of the United States. As one of only two contractors capable of building nuclear-powered submarines, Electric Boat was an indispensable pillar of "the U.S. submarine production base." [44] Even Senators Edward Kennedy and John Kerry, staunch liberals from Massachusetts, joined the chorus of Seawolf supporters, perhaps because Raytheon Corporation in their home state was a major subcontractor on the project. Congress approved the second Seawolf and provided funding to investigate the merits of a third boat. [45]

Bush accepted the second boat but not the third. Bill Clinton, while campaigning for the Democratic nomination in 1992, had promised to authorize another Seawolf if elected, but when he took office in 1993, he walked back that commitment. Eventually, however, he also bowed to the political pressure for a third boat, agreeing to a weapon system that the country did not want or need, what economist Ann Markusen has called a $2 billion "white elephant."[46] This was a classic case of the MIC forcing on the president military spending that he considered wasteful and unnecessary. The difference in this case was that the military sided with the commander-in-chief. This policy reversal was driven mostly by industry and Congress.[47]

Conversion, the second strategy for dealing with defense cutbacks, held little appeal for industry, but engaged those bent on rationalizing national security and promoting the welfare of the country as a whole.[48] The idea was to convert surplus manufacturing capacity in the DIB to production of commercial goods—from tanks to bulldozers, for example, or swords to plowshares. It could theoretically rescue failing defense contractors while stimulating the overall economy. In the post-Vietnam era of the 1980s, when President Reagan's defense buildup was adding to the national debt, many economists saw defense spending as a "burden."[49] Even when focusing on the effects of military research and development (R&D), including spin-offs, economists worried about opportunity costs and diversion of scientific and engineering talent from civilian pursuits.[50] Some scholars argued that defense spending had a positive impact, but they were in the minority.[51]

The defense industry resisted conversion for several reasons. First, it was easier said than done. Many military contractors worked primarily in the defense marketplace, with its highly specialized ways of doing business. Such companies understood the politics and mechanics of government contracting and how to satisfy a customer that valued quality over cost. Defense contractors often had little corporate experience in marketing and cost control, and they knew little of marketplace competition based on price and customer preferences.

There were sterling examples of conversion in the early 1990s. Westinghouse's division of defense electronic systems, for example, expanded its focus to include airline reservations systems, mail-sorting equipment, tracking systems for mass transit and fleets of commercial vehicles, and weather radar. Lockheed contracted for commercial airline maintenance, collection of parking fines, and design of small satellites for cellular phone networks. Martin Marietta took up management of computer systems for the Department of Housing and Urban Development and produced commercial aircraft components and mail-sorting systems.

Conversion appealed most to reformers such as Ann Markusen and Jacques Gansler, who focused on the *national* interest. Like Dwight Eisenhower, they believed that a robust national economy was the sine qua non of national security. However, by definition, that was not the goal of the MIC, which focused on the institutional interests of its constituents. It was possible for both sides to seek out win-win conversions, in which defense contractors found profitable options in the commercial marketplace to which their skills, capital plant, and personnel could be redirected. But conversion proved easier—and more attractive—in theory than in practice.[52]

The third tactic of defense contractors in the face of steeply declining defense budgets, consolidation, was related to the other two: companies in the DIB might merge to consolidate their portfolios, minimize competition, reduce costs, and muster sufficient assets to weather the drought. Indeed, the new Clinton administration actually recommended this to the industry. In 1993, Deputy Secretary of Defense William Perry, who had executed the "second offset" in the Carter administration, convened a dinner meeting of executives from the country's major defense contractors. He announced that they should not expect the decline in defense spending to abate. Perry warned that perhaps half the companies present would cease to exist in the near future. He recommended consolidation, signaling that the Clinton administration would go easy on merger proposals within the DIB, weighing national security above monopoly concerns. The message was so stark and draconian that the dinner came to be called "the last supper."[53]

In the resulting "fire sale," as one industry analyst described it, a riot of mergers and acquisitions transformed the DIB, eliminating some companies altogether and assimilating others in new conglomerations over the course of the 1990s.[54] Remarkably, the top tier of defense contractors nonetheless looked very much the same at the end of the decade, though overall numbers were significantly reduced. A comparison of DIB leaders in 1992 and 2000 shows how consolidation actually masked the inertia of the old order (table 1).

The movements of these companies suggest some consequences of consolidation in the 1990s. McDonnell Douglas, itself the result of a 1967 merger of two major aircraft manufacturers, merged with Boeing in 1997 to move Boeing on the list of top defense contractors from ninth in 1992 to second in 2000. The

TABLE I. Leading Defense Contractors

1992	2000
McDonnell Douglas	Lockheed Martin
Northrop	Boeing
Lockheed	Raytheon
General Dynamics	General Dynamics
General Electric	Northrop Grumman

merger also configured Boeing to continue as a dominant player, because it gave the new company prominent positions in both military and civilian aerospace realms. Northrop bought Grumman Aerospace in 1994 to diversify from its previously narrow focus on military aircraft, adding still more companies in the course of the 1990s to become a major integrator. General Dynamics reversed course on its initial sell-off of assets to become a major integrator by buying up other companies falling by the wayside. General Electric, under the guidance of its new CEO, the legendary Jack Welch, sold off its aerospace business to Martin Marietta, while retaining its jet engine division, which produced both military and commercial models. It fell to fifth among defense contractors in 2005, but the parent company rose to be the largest corporation in the world in 2000, though no longer among the top five defense contractors. Finally, Raytheon (cofounded in 1922 by World War II science icon Vannevar Bush) rose from being the fifth defense contractor in 1991 to third in 2000 by moving aggressively into defense electronics, aviation, and missiles, including the important Patriot anti-missile missile.

Finally, Lockheed merged with Martin Marietta (itself a

product of an earlier merger) in a story that warrants its own telling. Even by the standards suggested at the last supper, this was an exceptional consolidation of assets and power. It was made possible in part by the deft political maneuvering and unparalleled influence of Norman Augustine, the CEO of Martin Marietta. Not only did Augustine sell the merger to the political establishment in Washington, he even parlayed a government offer to subsidize such mergers into "restructuring costs" that paid for plant closings, relocations, severance pay, and "golden parachutes" for board members of the two firms who lost their positions.[55] Augustine himself received $8.2 million in the settlement and emerged as the CEO of the new Lockheed Martin company, the country's largest defense contractor at birth in 1995. Senator Bernie Sanders (D, NH) labeled this consolidation, with its attendant job losses, "payoffs for layoffs." After further enlarging Lockheed Martin with other mergers and acquisitions, Augustine, now chairman of the board, tried to buy aerospace giant Northrop Grumman in 1997. The Justice Department rejected the deal unless Lockheed Martin agreed to sell off Northrop Grumman's electronics assets, fearing a vertical integration in some fields of aerospace manufacture.[56] The DoD sided with the Justice Department and Lockheed Martin abandoned the effort.

The rejection of the Lockheed Martin/Northrop Grumman merger marked the climax of what Jacques Gansler has called the "consolidation orgy" of the Clinton administration, when the number of leading defense contractors fell from 36 to eight and spending on acquisitions and mergers rose from $2.7 billion to $31.2 billion.[57] The results were certainly epochal,[58] but they weakened a guiding tenet of the DIB throughout the Cold War:

sustain multiple industrial sources for critical weapons and equipment, both to promote competition and to ensure backup of critical manufacturing capabilities. Put not all your eggs in a few—let alone one—baskets.

Research and Development

As the defense industrial base reorganized itself for the new world order, it also continued the transformation of research and development (R&D) that had been going on in the country at large since the earliest days of the MIC.[59] The large industrial research laboratories of the mid-twentieth century—American Telephone and Telegraph (AT&T), International Business Machines (IBM), the Radio Corporation of America (RCA), Westinghouse, and General Electric, to name a few—produced Nobel prizewinning research and fueled the American economy with world-beating innovations.[60] In time, however, the large corporations funding these research laboratories sought less expensive ways to innovate. They shrank their laboratories, focused them more narrowly on applied research directed toward immediate product applications, outsourced specific research problems to universities and specialized laboratories, and acquired companies that already had the answers they needed.

The decline of industrial research laboratories did not, however, signal a lessening of the American commitment to research and development prescribed by Vannevar Bush at the end of World War II. Both government and business spending on R&D accelerated through the Cold War, accounting for 90% of US R&D funding since 1953. The government funded most of that

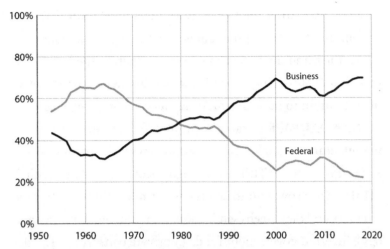

CHART 3. Percentages of US R&D expenditures by federal government and business sector, 1953–2018. *National Science Foundation, "National Patterns of R&D Resources: 2017–2018 Data Update," NSF 20-307, January 8, 2020, table 6.*

research until 1980, when business investment began to increasingly outpace federal spending (chart 3).[81]

By 2000, business accounted for 69.4% of US R&D spending, compared with 25.1% for the federal government. Almost from the beginning of the Cold War, business investment in R&D had risen steadily, while the federal investment declined as a percentage of GDP and of total national spending on R&D.[82] Military R&D, always the largest component of federal R&D, followed overall military spending—moving up and down with changing administrations but always trending generally lower as a percentage of GDP and the federal budget. Over time, it also became a smaller part of government R&D spending. In 1953, for example, the DoD accounted for 84% of federal R&D; that number fell to 54% in 1999.[83] Like the government as a whole, the DoD ceded national leadership in R&D spending to the private sector, casting doubt on the services' post–World War II commitment to world preeminence in military technology.

At the same time, the armed forces experimented with new models for sponsoring research and development, paying particular attention to rapid Japanese advances in fields such as computers.[64] DARPA, for example, promoted computer design and manufacture through creative ventures with the private sector. Programs such as VHSIC (very high speed integrated circuits), MIMIC (microwave/millimeter wave monolithic integrated circuits), and Sematech (semiconductor manufacturing technology), invested government funds in private consortia.[65] The defense establishment had to tread carefully in this area, lest its sponsorships step over the line into "national industrial policy," picking winners and losers and upsetting the natural balance of the capitalist marketplace with the heavy weight of the government purse. DARPA Director Craig Fields was sacked in 1990 when he promoted and then exploited new legislation to allow government funds to be invested in private corporations working on projects of potential benefit to the country. For example, he put $4 million of DARPA funding in Gazelle Microcircuits, Inc., which was doing early work on gallium arsenide computer chips, a technology of great military potential. But when he invested in research on high-definition television (HDTV), for similar reasons, he was fired by President George H. W. Bush, for crossing a line into the commercial sector.[66] Republicans were more likely than Democrats to call out such government behavior, but they were also most likely to support military programs that strengthened national security. Indeed, many critics claimed that government nurturing of the DIB was itself a national industrial policy.[67] So even when Republicans controlled the White House or one or both houses of Congress, radical experiments in funding defense R&D might be tolerated.

A watershed was reached in 1999, when the CIA, a bona fide member of the national security establishment, provided seed money for In-Q-Tel, a public-private, venture-capital firm designed to invest in new companies developing promising ideas and technologies of potential usefulness to the US intelligence community.[68] The CIA established the purpose and ground rules for the firm and retained the right to choose its board members and principal executives. But those officers were nonetheless given considerable leeway in identifying start-ups and other small companies with promising ideas and business plans. In-Q-Tel could trace its lineage to Sematech and other computer and microelectronics partnerships of the 1980s and 1990s, but it attracted positive public attention by actually investing capital in a public company and relying on the commercial success of that company to recover its investment. Such returns were recycled.[69]

Not surprisingly, In-Q-Tel often sought investments in dual-use technologies, or even in commercial-off-the-shelf (COTS) technologies. This preference reflected two growing patterns in military procurement. Technologies in the "microelectronics domain" were often emerging from commercial sources faster than the military could identify, nurture, and purchase innovations through its own contracting. Commercial technologies like cell phones, Wi-Fi, tablets, even GPS (a military technology soon shared with the public) were advancing capabilities faster than the military could keep up. It was quicker and cheaper to buy those products COTS than to develop the bespoke, highly specialized versions that the armed services in a previous day would have custom ordered. It only made sense to identify and adopt dual-use technologies that were appearing in the commercial marketplace. And if a company had a new and promising idea that might have both military and commercial applications, why not help it bring the product to market? The government might even get financial return on its investment to redeploy elsewhere. And of course, combined public-private investments reinforced the argument that Nora Simon and Alain Minc had made for a "thick fabric" of American institutions supporting computer development in the 1980s.[70] It also confirmed the 1973 prediction of Daniel Bell that the United States was moving toward a post-industrial society in which the early Cold War model of industrial research by large corporations would give way to more government-industry collaboration and science-based innovation. Fred Block and Matthew W. Keller confirmed that just such a transformation took place in the decades after 1970, turning the United States into a "developmental network state" very similar to what Weiss called a "national technology enterprise."[71]

Even with these innovations filtering into the DIB and changing some of its ways of doing business, many of the classic hallmarks of the Cold War Military-Industrial Complex remained in evidence. As the DIB shrank through consolidation, leaving a smaller number of larger companies, the imperative to retain multiple competitors in each field of endeavor became more difficult. Northrop got the B-2 bomber to keep it in business, and Lockheed, the largest and richest of all defense contractors, got the F-35 by the same distributive logic. Ironically, Lockheed had produced the too-costly-to-fly F-22, whose shortcomings called forth the even more expensive F-35. Boeing repeatedly lost follow-ons to base contracts it held so that other less flush companies could share some of the wealth.[72]

All the while, the DIB continued to lobby Congress, to push for a stronger American military, to protest contract cancellations, and to bemoan what it perceived to be the narrowing gap between America's arsenal and those of its potential adversaries.

The revolving door between government employment and private positions in companies of the DIB turned with undiminished speed. The collusion at work was exposed with damning clarity on more than one occasion. For example, Darlene Druyun, the air force's principal negotiator for tanker aircraft contracts, discussed a post-government job with Boeing while she was negotiating with the company a $23.5 billion contract to lease tanker aircraft.[73] William Black, after 40 years at the National Security Agency, joined Science Applications International Corporation (SAIC) in 1997, "for the sole purpose of soliciting NSA business." When SAIC won a preliminary contract for Project Trailblazer, to update American equipment for eavesdropping on the Russians, Black returned to the NSA as deputy director. A year later, the NSA awarded the master contract for Trailblazer to SAIC. Black was just one of many officials from the military and intelligence communities to join SAIC after leaving government service.[74]

Since the Cold War, contracting has adapted to changing mechanisms and institutions of research and development and procurement.[75] For example, the exceptional tool of "Other Transaction Authority," granted to NASA in 1958 at the outset of the space race with the Soviet Union, was extended to the DoD in 1989.[76] It allowed a few agencies to use non-procurement contracts for R&D and development of prototypes. The authority was further extended after 9/11 to other agencies, including Homeland Security. Other non-procurement contracts for public-private partnerships and intergovernmental agreements were also available to agencies in the security establishment.[77] In general, these exceptional contracts loosened government controls on contracting processes.

Neither structural changes nor cycles of reform, however, could meliorate the chronic flaws in government procurement of goods, services, and R&D. As J. Ronald Fox makes clear in his 2011 official history of DoD acquisition reform, the "system has been strongly resistant to change." Fox confirms a 1992 GAO conclusion that the DoD "acquisition culture" has persistently erected "formidable barriers to acquisition reform."[78]

The twin curses of buying-in and gold-plating continued to haunt defense contracting. Companies knew that if they once won a contract by bidding low, they could renegotiate later to make up the shortfall. They might, for example, overcharge for client-initiated changes (gold-plating), increase funding for unanticipated problems, overcharge for supplies and equipment not specified in the base contract ($435 hammers and $1,280 coffee cup warmers), and seek reimbursement for delays beyond the control of the contractor. By the time such add-ons arose, the principal would be so committed to the project and in need of its successful completion that it could not cancel and start over with another contractor. Furthermore, contractors were allowed to recover losses from a first production run in subsequent runs. So even if they had lowballed the initial bid, they could write their losses into a follow-on contract, for which they would be the most advantaged bidder.

All the while, government officials overseeing a contract often had a vested interest in making the contract work and in ensuring the continuing viability of the contractor to complete the deal. For example, the air force helped Lockheed cover up and then bounce back from the C-5A scandal because it did not want to start over with a new contract. This misconduct prompted adoption of the "fly-before-you-buy" concept, calling for demonstration flights before commitment to production and procurement. Proposed by Deputy Secretary of Defense David Packard, cofounder of information technology mega-firm Hewlett-Packard, the policy was embraced by Defense Secretary Melvin Laird during the Nixon administration's consideration of the B-1 "born-again bomber."[79]

The DIB also promoted overseas arms sales. George H. W. Bush spent his last year as vice president and his entire term as president trying to distance himself from the Iran-Contra scandal of the Reagan administration. He repeatedly advocated reductions in all international arms sales by the United States, yet steadily increased the volume of those sales while in office.[80] Beginning in 1990, the United States replaced the Soviet Union as the world's leading arms supplier, consistently making about half the world's arms transfers through the Bush administration and most of the Clinton administration.[81] These transfers, of course, not only cemented relations with friendly countries but also satisfied the MIC, providing more sales for industry, cheaper unit prices for the DoD, and more military spending for Congress.

Chapter Nine

War on Terror

Transitioning to a War on Terror

As the 2000 election approached, the Clinton administration reversed its previously unrelenting cuts in defense spending. Democrats had reason to worry that American voters might find defense spending at just 2.9% of GDP too lean for comfort. Furthermore, President Clinton had opted for armed conflict at about the same rate as his Republican predecessors—in Bosnia, Kosovo, Somalia, Haiti, Kuwait, and Serbia. As his feisty Secretary of State Madeleine Albright had demanded of his circumspect Chairman of the Joint Chiefs of Staff Colin Powell, "What are you saving this superb military for, Colin, if we can't use it?"[1] Using it meant sustaining it.

So, the military budget rose slightly in 2000, and again in 2001, when George W. Bush assumed the presidency after an election so close that it was decided in the Supreme Court. His victory hardly provided a mandate for further increases in defense spending, though he surely hoped to achieve them if he could. Two priorities distinguished the incoming George W. Bush administration from standard Republican orthodoxy. First, he resolved to energize the quest for ballistic missile defense. This goal paid tribute to President Reagan and his Strategic Defense Initiative of 1983, seeking to galvanize a troubled program on which the Military-Industrial Complex had failed to deliver. The president sought to deploy a workable system during his tenure in office.

The second clear objective was defense reform, the pet concern of his Secretary of Defense Donald Rumsfeld. Along with new Vice President Dick Cheney, Donald Rumsfeld was coleader of the self-described "Vulcans," an informal group of high-powered Republicans favoring a stronger military establishment and a more assertive foreign policy. Many were veterans of the George H. W. Bush administration. Rumsfeld had served as secretary of defense in the Gerald Ford administration and as a White House advisor, director of the CIA, and aide to Secretary of Defense Cheney in the first Bush administration. After stints in the private sector and several passes through the revolving door, the two friends and allies were back in government together and more powerful than ever. While Cheney had served as CEO of Halliburton, a multinational oil field service company and a top-10 defense contractor in the 1990–1991 Iraq war, Rumsfeld had led three commercial firms to financial success and accepted several appointments in public service in the Reagan, Bush I, and Clinton administrations. He also served on the Project for the New American Century (PNAC) and in other lobbying and advisory roles. He was, in short, an experienced and influential defense intellectual, a pillar of the Republican establishment, and a seasoned Washington insider.

Rumsfeld entered the Pentagon determined to transform the

military establishment and redirect its focus, both goals with major implications for the Military-Industrial Complex. While he stopped short of embracing the Revolution in Military Affairs (RMA) and net-centric warfare, he did recognize the need to develop "advanced conventional capabilities."[2] He believed that the services, especially the army, were locked in a Cold War paradigm, still planning, training, and buying weapons for a conventional war with Russia on the plains of Europe. Emblematic of his frustration was the Crusader self-propelled howitzer. This 43-ton, 155-mm gun, originally part of a Reagan-era program to modernize all army armor in the depths of the Cold War, was designed to replace the Paladin howitzer, which had entered service in 1963.[3] Rumsfeld's skepticism about the Crusader quickly drew him into a classic tug of war with the MIC. It mattered little that the General Accounting Office had recommended that the army either upgrade the Paladin or buy the comparable German Panzerhaubitze 2000.[4] Virtually all facets of the MIC entered the fray over the Crusader. The army had gold-plated the gun with features such as a cooled cannon for higher rates of fire, automatic handling of munitions, composite armor, and other weighty, expensive, and unproven technologies. By 2002, price escalation had driven the cost to $11 billion for 480 vehicles. The water-cooled cannon had already been dropped as unworkable. Some army personnel, without authorization, briefed friendly congressmen on the need to save the program. Manufacturer United Defense joined the "furious lobbying" to block cancellation.[5] A Pentagon staffer called the Crusader "a wonderful system—for a legacy world."[6] President Bush sided with his older, experienced secretary of defense, signaling a willingness to defy the MIC.

The issue hung in the balance when the terrorist attacks of 9/11 changed the calculus of US national security policy. The shock of almost 3,000 civilian fatalities on American soil, the greatest such toll since Pearl Harbor, shattered the hubris behind the "new world order" and suggested that nonstate terrorists posed a greater threat to US security than any would-be peer rival. But the same remedy might apply to both challenges. A major strengthening of the US military establishment seemed obviously in order and politically attainable. And a military buildup could enable the United States to exact retribution on those responsible for 9/11 while simultaneously funding the armed forces necessary to deter or preempt the emergence of peer rivals.[7] The Crusader disappeared from President's Bush's next military budget.

President Bush announced two pivotal policy decisions. First, the United States declared a "War on Terror" or "Global War on Terrorism," vague, but stirring, epithets that captured the direction in which Bush proposed to lead the country. Second, the president pronounced that the United States would pursue the terrorists to the ends of the Earth, rooting them out in their lairs—preemptively—rather than waiting passively for them to attack the homeland again. The principle at work was classic: a good offense is the best defense. Of course, improvements in "homeland security" complemented the Bush strategy in the coming months and years. However, the War on Terror meant that the Department of Defense would focus on conventional "force projection," beginning in the Middle East and southwest Asia, where the 9/11 attacks had originated.

For its part, the MIC could expect orders for increased production of the arms and equipment already in hand. Unlike the Reagan defense buildup of the early 1980s, Bush's call for increased defense spending did not prioritize research and development, the backbone of American military dominance. As Secretary of Defense Rumsfeld put it, with characteristic disdain for those service personnel still fighting the last war, "You go to war with the army you've got." There was no time to reform the army or equip it with the arsenal best suited to the immediate challenge—as Rumsfeld had been recommending for years. The country would have to rely on the most promising way of war currently in harness—the Revolution in Military Affairs. Rumsfeld certainly appreciated this movement, though he wished that the military services, especially the conservative army, had advanced the agenda more imaginatively in the 1990s. Funding for the Crusader migrated quietly to the army's "Future Combat Systems," an even more ambitious program that would fail even more disastrously in the fullness of time.

At the same time, the Bush administration reorganized the intelligence community (IC) in response to a widespread perception that 9/11 was an intelligence failure. The attacks of 9/11 exposed an intelligence establishment riven by institutional parochialism, civil-military discord, interservice rivalry, bureaucratic infighting, and the politicization of intelligence.[8] The 17 agencies comprising the new IC were divided between a National Intelligence Program (NIP) under a director of national intelligence (DNI) and a Military Intelligence Program (MIP), soon directed by the new office of under secretary of defense for intelligence. Secretary Rumsfeld, an accomplished bureaucratic infighter, captured the key agencies for data collection and analysis: the National Security Agency (NSA), the National Reconnaissance Office (NRO), and the National Geospatial-Intelligence Agency (NGA), which uses global positioning and other related technologies for mapping, targeting, navigation, and other activities. The director of national intelligence, a cabinet-level official, would nominally lead the entire IC and command a larger intelligence budget than the secretary of defense. Furthermore, he replaced the CIA director in daily briefing of the president. Still, the reorganization of the intelligence community did not remove all the interagency competition that had handicapped the IC before 9/11.

In addition to expanding and reorganizing the intelligence community, the Bush Administration also expanded some existing agencies, such as the National Security Agency (within the MIP), which increased its international surveillance activities and even collected some domestic communications under authority of the Patriot Act (October 26, 2001). This last change gave the government unprecedented access to the communications of American citizens.[9] These drastic measures reminded some observers of Harold Lasswell's warnings of a "garrison state," which had so concerned President Eisenhower. In time, the draconian measures taken in the first months and years of the War on Terror would be moderated as the memory of 9/11 faded and the intrusions on privacy became manifest. In 2001 and 2002, however, Congress and the American people gave the government extraordinary powers to remake the national security establishment.

The last piece of the post-9/11 organizational restructuring of government was the creation of a new cabinet-level Department of Homeland Security (DHS), weaving together some existing government agencies—Coast Guard, Secret Service, Federal Protective Service, and so on—with new entities such as the Transportation Security Administration (TSA). The DHS employs almost 100,000 civil servants and more than 40,000 contractors, making it the third largest organization of the federal government, behind only DoD and Veterans Affairs. Thus, the three largest agencies of government serve what some critics call the National Security State, raising new questions about what should count as military or defense spending.[10] The DHS's 2018 budget of almost $48 billion was smaller than that of the intelligence community but larger than the marine corps. Most of its costs are personnel, to screen air passengers, monitor borders and ports of entry, and cooperate with agencies of foreign and domestic intelligence to stop security threats from entering the country. Like the intelligence community—but unlike the Department of Defense—it spends most of its contracting money on procuring services, not materials. This trend highlights one of the major changes to national security contracting since the attacks of 9/11.

The RMA Goes to War

The Bush administration's first overseas campaign of the War on Terror boded well for the government's strategy and for the RMA. In a proof-of-concept campaign in Afghanistan in 2001, the US military sent a small number of special operations forces into the country to hunt down Osama bin Laden, the Al-Qaeda leader who had directed the 9/11 attacks, and the Taliban extremists who had taken over Afghanistan and given bin Laden and his fellow jihadists safe haven. American soldiers and CIA operatives on the ground employed "net-centric warfare" to coordinate their operations; track enemy communications and movements; communicate with their superiors; support ground forces, intelligence sources, and weapons delivery systems; visually fix their targets night or day; and call in air strikes by US Navy and Air Force tactical aircraft and unmanned aerial vehicles. Within a matter of weeks, at the cost of just seven American lives, a handful of boots on the ground, working with Afghanistani forces, drove the Taliban from power and Al-Qaeda into safe haven in the mountains on the border with Pakistan. By the end of 2001, only human error had prevented the American forces from trapping and destroying the fleeing Al-Qaeda leadership. One possible lesson was that the legacy military establishment from the Cold War, armed and equipped by the MIC, had produced a more effective fighting force than Rumsfeld had imagined. Or perhaps, as its advocates had claimed, the RMA had equipped that traditional fighting force with a brace of new high-technology arms and equipment that elevated its capabilities to a virtually invincible level.[11]

Buoyed by the stunning success in Afghanistan, the Bush administration raised its sights. In the first week after 9/11, Paul Wolfowitz advocated "ending states who sponsor terrorism" and began planning military action against Iraq.[12] Without any evidence yet that Saddam Hussein had played a role in the 9/11 attacks, the Bush team turned its attention to the larger Vulcan agenda: rolling up those other states beyond Afghanistan (North Korea, Iran, and Iraq) that might harbor ambitions or terrorists hostile to the United States. As this confrontational intent became manifest, opposition arose in the United States and among its allies. Unwilling, and perhaps unable, to proceed alone, the Bush administration offered Americans and allies a variety of justifications. None of them gained purchase until it was asserted by the US intelligence community that Iraq's tyrannical ruler, Saddam Hussein, was secretly developing weapons of mass destruction—nuclear, chemical, and biological. The evidence for this claim proved too weak to carry the argument until Secretary of State Colin Powell endorsed the finding in testimony before the United Nations. With lukewarm support at home and abroad, the US led a coalition of ambivalent allies into Iraq on March 19, 2003. The invasion precipitated a war—really two wars—for which American leaders, citizens, and their armed forces were unprepared.

At first, the invasion appeared to once more validate the RMA, and by extension the MIC that made it possible. US and allied forces rolled up an outmatched assemblage of Iraqi and allied fighters and captured Baghdad in little over a month. Importantly, only 138 US personnel died, compared with perhaps 10,000 Iraqi military and civilian deaths.[13] For champions of the RMA, this was a grisly but satisfying proof of concept. The victory was, as in Afghanistan, swift and decisive, sufficient for George W. Bush to appear on the deck of a US aircraft carrier on May 1, 2003, in front of a banner reading "Mission Accomplished." The MIC shared President Bush's appraisal of the invasion. Its arms and equipment had now enabled fighting forces to achieve two impressive triumphs in the wake of 9/11. Defense spending was rising, and the Vulcan agenda for a more assertive US grand strategy seemed ascendant.

Behind the scenes, more veterans of the defense industry had passed through the revolving door between the government and private sectors. I. Lewis "Scooter" Libby, for example, the actual author of the "no peer rival" policy in the George H. W. Bush administration and a former consultant for Northrop Grumman, was now chief of staff for Vice President Cheney. Paul Wolfowitz, Libby's former instructor at Yale and his boss when he oversaw drafting of the "no peer rival" policy, and as well a fellow associate with the PNAC, had not joined the defense industry while out of office, rather serving as dean of the Johns Hopkins School of Advanced International Study. From that post in Washington, he remained active in the PNAC and the emerging Vulcan club of foreign policy mandarins.[14] These and other Vulcans and like-minded hawks crafted and welcomed President Bush's famous "Axis of Evil" pronouncement in his State of the Union address of January 29, 2002, suggesting that other regime changes—that is, Iran and North Korea—might be in America's future.

But overthrowing a government with devastating military force turned out to be easier than occupation, nation-building, and counter-insurrection. Indeed, Iraqi insurgents were already introducing the most important response to American military dominance—the improvised explosive device (IED). This brilliant innovation, as its name suggests, used simple, inexpensive, and readily available materials to construct mines for the roads and byways of Iraq. These could be exploded by contact, com-

FIGURE 8. This IED (improvised explosive device) was discovered by Iraqi police in eastern Baghdad on November 7, 2005. Four artillery shells were wired to an anti-tank mine to be detonated simultaneously. *Licensed under CC0, public domain.*

mand, or timer, depending on the target and the circumstance. The explosive material could be salvaged from unexploded American ordnance. The detonators could include cell phones, pressure mechanisms, manual switches, or other similar devices (figure 8). American occupation forces moving about the countryside began suffering fearsome casualties that included death, bodily trauma, brain injuries from concussion, and severe psychological stress. Mine-sweeping technologies were too scarce and too slow to meet demand and only tanks offered robust protection. The RMA, prepared as it was for conventional warfare on the plains of Europe, had no ready answer to the IED.

The military establishment took some immediate steps to change its tactics for counterinsurgency operations, but its operational doctrine was based on vehicular movement to service a long, heavy logistical tail. The RMA was powerful, but it was also resource intensive—computers and communications equipment, generators, air conditioners, runways and helipads, support personnel to man the equipment and process the data, and, perhaps above all, fuel and ammunition. All these people and materials moved primarily by road. Not surprisingly, the enemy targeted those roads and the Americans bound to them. In 2003, beginning in June, IED attacks jumped from 22 to more than 600 a month. Two years later, they had reached more than 2,000 a month, as many as 100 in a single day.[15] By early 2007, more than 3,300 American military personnel had been killed in Iraq, as many as 70% of them by IEDs. It was said that the enemy had gotten "inside the [Marine] Corps OODA loop," the ultimate humiliation of the RMA.[16] When efforts to avoid and interdict IEDs failed and "up-armoring" of existing vehicles proved ineffective, the military finally turned in desperation to the MIC for a technological fix.

The US Army and Marine Corps already had at their disposal mine-protected vehicles, used for decades for explosive ordnance disposal and other combat-engineering missions, but they proved inadequate against some of the huge IEDs being deployed in Iraq.[17] ARPA had studied IEDs in 1971, concluding that there was no good response to them.[18] Beginning early in 2006, DoD circulated among many potential contractors specifications for three kinds of MRAPs—Mine-Resistant Ambush Protected vehicles. By November, preliminary contracts were let to nine bidders for two of the three possible vehicular types then envisioned by DoD. Ten production MRAPs were accepted in February 2007, increasing to 1,160 in December. The rising tide of IED casualties began to subside.

MRAPs were a remarkable achievement, an ad hoc combat

system invented on the fly and rushed into production by multiple manufacturers.[19] Less than 18 months after publication of an "initial needs statement," the first MRAP appeared in the war zone.[20] By the time the program ended, it had spent $47.4 billion with as many as seven different contractors producing almost 28,000 vehicles, of which 24,095 were deployed in Iraq or Afghanistan. Defense Secretary Robert Gates, who made MRAP the DoD's number one procurement priority in May 2007, would later claim that MRAP was "the first major defense procurement program [since World War II] to go from concept to full-scale production in less than a year."[21] Just as the IED symbolized the indigenous, asymmetric answer to the Revolution in Military Affairs, the MRAP became America's "new icon of Operation Iraqi Freedom" (figure 9).[22]

The MRAP story was a microcosm of the MIC early in the twenty-first century. The Pentagon responded slowly to the IED, trying to tweak existing assets when confronted with an unexpected threat.[23] Then it initiated a crash program, funded parallel developments by multiple contractors, specified sophisticated armament (machine guns, grenade launchers, missiles) and capabilities (300-mile range, 65 mph on improved roads, sophisticated sensors and communications), waived standard procedures for development and testing, and rushed untested vehicles into combat at an average cost of $1.7 million per vehicle.[24] Not surprisingly, the vehicles experienced operating problems, such as inadequate suspension, proclivity to rolling over, and limited off-road capability.[25] Still, the defense industry had produced, in short order, state-of-the-art vehicles that succeeded in their main mission: reducing casualties from IEDs. Casualties weak-

FIGURE 9. An Air Force MRAP Cougar, pictured here at the Aviation Museum at Robins Air Force Base, Georgia, in front of a B-1B Lancer, the "born-again bomber." This vehicle suffered heavy damage in an IED attack in Afghanistan in 2014, but none of its occupants were injured. *Robins Air Force Base, Georgia.*

ened public support for the war and, according to the military, cost far more than the MRAPs. One military study estimated that a single, conventional vehicle carrying one officer and five enlisted men represented $2.5 million in personnel replacement costs.[26] The Joint Program Office for the MRAP claimed that the vehicles saved 40,000 American lives in Iraq and Afghanistan by assuming that every enemy-initiated attack on an MRAP would have resulted in the deaths of all crew members in any other vehicle.[27] The defense industry, the military services, and congressional supporters all wanted to win the war for public opinion as well as the wars in Iraq and Afghanistan. Of course, the enemy adapted to MRAPs, and the IED remained a potent threat in Iraq and Afghanistan, but MRAPs helped to keep American casualties at a tolerable level.

The Intelligence-Industrial Complex

While the military services struggled to adapt to counterinsurgency in Iraq and Afghanistan, the intelligence community struggled to adapt to its restructuring between 2001 and 2005. The National Intelligence Program (NIP) now oversaw strategic intelligence affecting national strategy and multiple agencies, while the Military Intelligence Program (MIP) focused on tactical and operational intelligence of particular interest to the armed services. The result strengthened the hand of the secretary of defense in key areas of intelligence gathering, but it left the DNI with a budget that has been two-to-three times as large as the MIP budget since 2007.[28] In practice, many intelligence activities are funded jointly by the NIP and the MIP. These organizational arrangements make less difference to the contractors than they do to the bureaucratic rivals within the Intelligence Community.

Within that community, however, significant changes were accelerating in the wake of 9/11. Intelligence agencies, especially those working in collecting intelligence by technical means—as opposed to "open source intelligence" (osint) and "human intelligence" (humint)—needed similar technological assets: satellites, computers, communications equipment, sensors, data banks, data mining, and more.[29] As was true of those military services becoming ever more dependent on the microelectronics domain, agencies of the IC found themselves turning to the existing commercial tech industry, sometimes for commercial off-the-shelf (COTS) products and sometimes for directed, demand-pull research and development. And, thus, it happened that an

Intelligence-Industrial Complex (IIC) came into existence to service the country's newly perceived need for greatly enhanced technical means of intelligence gathering and processing.

Intelligence reform spurred increasing demand for public information about the scale of America's intelligence community. Throughout the Cold War and the New World Order and into the War on Terror, the US intelligence budget was a closely kept secret and a matter of educated guesswork by students of American national security. Most IC spending was hidden in obscure categories within the DoD budget.[30] After the 9/11 attacks and the reorganization of the IC, the "top line" budgets for the NIP and MIP were made public, beginning in 2007.[31] In that year, NIP spending was $43.5 billion, and the MIP was $20 billion. Combined, these amounted to about 11% of that year's total military budget of $551 billion. In general, spending has remained at about that proportion ever since.[32]

The IIC need for innovation has been no less urgent than that of the MIC, of which it was a part. Most of that innovation came from contractors and from research and development sponsored within the microelectronics domain by IC agencies and by other arms of the MIC, such as DARPA. In many fields, the commercial industry was already developing state-of-the art technologies—products such as supercomputers, computer chips, data storage and mining, cryptography and cryptanalysis (making and breaking codes), data transmission and interception, and data analysis. Contracting with the tech community for existing and aspirational technologies and services required some adaptation of existing contractual practice and some navigation of the boundaries between open and classified technologies.

In-Q-Tel offers one particularly revealing example of how the IIC operated in this rapidly changing environment.[33] When the CIA created In-Q-Tel in 1999, it envisioned a hybrid venture capitalism that might be called "venture innovation."[34] It was going to make high-risk / high-payoff investments in companies with promising ideas, not to make money but to develop capabilities of use to the American intelligence community. If the companies they supported also turned out to be commercially successful, so much the better. The CIA financial return on investment would allow it to reinvest in other promising companies, but the real goal was new intelligence technologies. The IC still funded some demand-pull research in companies, universities, and private research centers; but the low barriers to entry in the computer/electronics field, especially in software, meant that many promising ideas might appear in start-up companies trying to achieve proof of concept in the absence of traditional sources of capitalization. Government funding of such ventures flirts with national industrial policy—picking winners and losers—but it began on a comparatively small scale, without threatening major players who could bring pressure on Washington to stay out of the marketplace. The federal government had become more tolerant of the activities that had gotten DARPA Director Craig Fields fired in 1990.

Beginning with just $28.5 million in 1999, In-Q-Tel had, by 2006, networked with 200 venture capital firms and 100 labs and research organizations, reviewed 5,500 business plans, invested in more than 90 companies, and delivered more than 130 innovations to the intelligence community.[35] Stephanie O'Sullivan, the CIA's director for science and technology, explained in

2006, "We're getting the best minds that we can find out there to think about our problems, to think about our need and the technology curve. It's a human capital investment, and it pays off."[36] For example, in 2003, In-Q-Tel discovered a company called Keyhole, Inc., which had developed software called EarthView to convert two-dimensional satellite imaging into 3D. But the Mountainview, California, company had been unable to attract a corporate sponsor or venture capital. In cooperation with the National Geospatial Intelligence Agency, In-Q-Tel provided an infusion of cash. In just two weeks, 3D images were in the hands of American forces in Iraq. The following year, Google bought out In-Q-Tel's investment in Keyhole and turned the technology into Google Earth. As with the military's previous decision to make its global positioning system (GPS) technology available to the public, this one raised serious questions about government-funded technological secrets being given away for public use and commercial exploitation. This practice further eroded the barrier between public and private, secret and open, but no one denied that the United States was better off having the technology first and the luxury of deciding whether or not to share it.

As Linda Weiss makes clear in *America Inc.?*, In-Q-Tel was neither unprecedented nor unique. Beginning as early as 1958, military and civilian agencies had experimented with what she calls government-sponsored VC (venture capital) funds (GVFs).[37] The difference after 9/11 is that such undertakings became far more numerous and varied and far more popular in what she and others call the National Security State (NSS). Furthermore, the process has contributed to the blurring of lines between public and private, emphasizing the need to promote commercialization of developments so that the government can purchase products COTS instead of made-to-order. She believes that this has transformed spin-off to "spin-around." Instead of trying to justify military or intelligence R&D by technology transfer to the commercial realm—like GPS—the government can now sponsor developments with potential to move in either direction, enhancing both national security and the national economy. As Eisenhower had believed, these two national priorities went hand in hand.

The "spin-off paradigm" of the Cold War passed through many iterations on its way to Weiss's "spin-around." "Spin-on" and "spin-in" focused on technology transfers from the commercial sector to the military, often achieved through COTS. "Spillover" drew attention to the ways in which innovations came to permeate all sectors regardless of their origins. Its opposite, "spin-away," addressed failures of technology transfer, often bred by secrecy, culture, and the alienation that sometimes divided civilian and military realms. What all of these characterizations shared in common was an appreciation of the changing paradigm for the transfer of technology between the military and civilian sectors. No longer was defense R&D the primary engine of American technological innovation.[38]

Whether the IIC is viewed as a subset of the MIC or a separate phenomenon, it nonetheless has many of the qualities of the Cold War institution. In the first two decades of the twenty-first century, both complexes engaged in mergers and acquisitions, resulting in a handful of dominant firms and a bewildering array of specialty enterprises that often serve as subcontractors. CACI (formerly the Consolidated Analysis Center, Inc.), for example, began as a software company in 1962 and then moved into other fields of computer development.[39] Between 2003 and 2019, it bought 33 companies, becoming in the process one of the top five contractors in the IIC. Increasingly, such firms provide services as well as material products, a phenomenon that will be discussed further below. Firms in both complexes hire retired government employees—military and civilian—both for their management experience and their connections with government officials who oversee contracts. The IIC also has its own trade organizations, such as the United States Geospatial Intelligence Foundation, which hosts an annual meeting that journalist Tim Shorrock describes as the "premier showcase for intelligence contractors and agencies alike."[40] The IIC lobbies both executive branch agencies and Congress, using jobs as a major argument for support from individual legislators. And it has iconic figures, such as John Michael McConnell, whose career tracked with the growth of the IC. A naval flag officer, he rose to prominence during the first Gulf War, when he served as chief intelligence officer for Chairman of the Joint Chiefs of Staff Colin Powell. He became director of the National Security Agency in 1992 before retiring as a vice admiral in 1996, whereupon he joined Booz Allen Hamilton, a perennial top-tier contractor for the intelligence community. In 2007, he went back through the revolving door to serve two years as director of national intelligence, and then back again to Booz Allen Hamilton, where he rose to vice chairman.[41]

But the IIC also differed in significant ways from the MIC. It often spoke of its relationship with industry as a "public-private partnership," stressing commonalities of interest and downplaying the formalities normally required between agent and client.[42] In addition to being an order of magnitude smaller, it was also more secretive. Even after its top-line budget was made public beginning in 2007, the specifics of its spending and contracts escaped the public scrutiny to which the rest of the military budget was subjected. It contracted for products focused more on dual-use technologies, especially in the microelectronics domain, and largely for that reason it entered more public-private ventures and bought more COTS products. The DNI has less control over the agencies in the NIP than the secretary of defense has over those in the MIP, though both enjoy the special access to the president entailed in holding a "cabinet-level" appointment.

Perhaps most importantly, the intelligence community may be said to have experienced more success than the defense community since 9/11. While the military struggled with intractable wars in Iraq and Afghanistan, the intelligence community shared with the new Department of Homeland Security much of the credit for discovering and intercepting all foreign attacks on the United States even remotely comparable to 9/11.[43]

Finally, it should be noted that the boundaries between the MIC and the IIC were porous. Both complexes had intelligence missions, and both engaged in overseas operations. The CIA, for example, continued to run covert operations and, for many years, to operate armed drones in hot spots around the world. Furthermore, some corporate consolidation entailed merging military and intelligence functions. In 2016, for example, Leidos Holdings, a leading contractor for the DoD and the NSA, merged with the information systems and global solutions division of Lockheed, becoming the top contractor for the IIC. In the process,

Leidos dislodged Lockheed from the position of number one IT contractor. Then, in 2018, MIC giant General Dynamics, of Seawolf fame, acquired CSRA, briefly the top IIC contractor, displacing Leidos in first place among all government IT contractors.[44] Both moves were made to strengthen the information technology portfolio of the purchasing company, thereby enhancing their integrator functions and opening up new contracting opportunities.[45] Acquisitions and mergers, accelerated by the "last supper" early in the Clinton administration, continue to consolidate both the MIC and IIC two decades into the twenty-first century.

Outsourcing

In the twenty-first century, government contracting for advanced weapon systems continued to follow patterns worked out during the Cold War. Some incremental reforms addressed past abuses—such as buying-in, capability greed, gold-plating, sole-sourcing, and bailing out—but those abuses are inherent in a marketplace that is simultaneously monopsonic and oligopolistic. Many observers of the contracting system in the twenty-first century have suggested that the problem resides in lack of enforcement of existing policies by knowledgeable and experienced government administrators.[46] Consolidation in the defense and intelligence communities has made it more difficult than ever to get multiple competing bids for some large projects. Allowing companies to partner with competitors when submitting bids can bring more talent and resources to bear on some projects, but it also poses a risk of incompatibility and rivalry between the partners. For example, Electric Boat and Newport News Shipbuilding partnered on the original contract for 29 Seawolf submarines but found that their production methods were incompatible.

The biggest change in MIC contracting after 9/11 was vastly increased contracts for services. This development reflected the high cost of personnel in the military and civil service. Indeed, the single cost category that distinguishes the US military from others around the world—especially China and Russia—is personnel costs. In 2008, for example, DoD's per person budget authority peaked at more than $450,000 for each of the 1.4 million active duty personnel in the American military.[47] Many factors are at work. First, wages in the United States are generally higher than those in other countries, meaning that the government must pay higher wages to attract qualified personnel. Second, the move to an all-volunteer armed force after the Vietnam War meant still higher base salaries and benefits to attract and retain personnel not facing conscription. This constraint was relieved somewhat by admission of more women into the American military, the loosening of constraints on sexual orientation, and the opening of most combat roles to women. Increasing costs of health care also strained the military budget, compounded by long-term medical care for increasing instances of PTSD resulting from IEDs and other hazards of the anti-insurrectionary wars in Iraq and Afghanistan. The cost of veterans' benefits for life rose with life expectancy, and bonuses for specialized personnel, such as pilots of high-performance aircraft, rose faster than base salaries.

Peter Singer published a groundbreaking analysis of military contracting for services in 2003. His *Corporate Warriors* offered a taxonomy of privatized military firms (PMFs), an international phenomenon in which the United States served as both principal and agent; that is, US corporations have provided such services internationally and US agencies have been major customers. Singer depicts PMFs as operating in three realms: direction, consultation, and support. Contractors may assist host countries in training, overseeing, and even leading the host's combatants. Consultants may assist their hosts by writing doctrine, organizing forces, and even planning campaigns and operations. Supporters provide services such as logistics, communications, security, and even public relations. In the twenty-first century, the international marketplace for PMFs has become so densely populated with contractors that states and even nonstate actors can hire boutique packages of military goods and services for ad hoc operations or long-term capabilities beyond the wherewithal of the principal. States like the US usually hire American PMFs, in the expectation that their goods and services will naturally align with American interests, though there is nothing in the contractual relationship that necessarily requires the agent to privilege its principal's goals over its own corporate priorities and its contractual obligations. In other words, the loyalty of such individual contractors may be to their own employer, not necessarily to the state that hired their service.

Singer notes that some PMFs have performed admirably in recent decades, providing goods and services that their principals could not duplicate and achieving their objectives efficiently and professionally. Halliburton and its former subsidiary, Brown and Root Services, held for decades the lion's share of the US Army's LOGCAP (Logistical Civil Augmentation Program) contracts to support the US Army in deployments in the Balkans, Iraq, Afghanistan, and elsewhere. By all accounts, they performed their assigned duties satisfactorily, but irregular contracting practices and lax oversight led to perennial accusations of fraud, overbilling, bribery, safety violations, money laundering, and even human trafficking.[48] In another case, in 1995, the chaos surrounding the breakup of Yugoslavia left Croatia and Bosnia locked in a stalemate with the more numerous and better equipped Serbs when an American PMF—Military Professional Resources Incorporated (MPRI)—under contract to the Croats sprang "Operation Storm" on the unsuspecting Serbs, ending the conflict in a matter of weeks and winning Croat independence.[49] "MPRI is exclusively made up of retired U.S. military personnel," notes Singer.[50] Many of the employees in such PMFs were former servicemen in the US military, who brought years of training and experience to their contracts. Whatever qualms state or nonstate actors might have had about outsourcing such work, they often achieved their goals sooner than they could have trying to build their own capabilities.

Such outsourcing contracts carried many of the same risks that attached to the Military-Industrial Complex. These include weak oversight, "cheating or intentional overcharging," and cutting corners, all without conclusive evidence that outsourcing necessarily saves money for the principal.[51] Furthermore, service contractors may walk away from a contract at any time, however much the principal may be relying upon them for security and mission achievement. Because both state and nonstate actors can hire PMFs, fundamental political theories of international security, based on the Weberian notion of war as organized armed conflict between states, slide into a new calculus of non-state actors as both principals and agents. Peter Singer concludes that "power is more fungible than ever," and that money and military capability can be converted back and forth by both state and nonstate actors. Criminal cartels have purchased military services to challenge local and even national law enforcement. By the same token, individual states or even the United Nations might hire private firms to aid law enforcement and even peace-keeping.[52]

Similarly, PMFs have disrupted traditional civil-military relations, outsourcing to private contractors functions usually performed by uniformed personnel, compromising the core value of civilian control of the military. Military personnel can become demoralized when working with contractors paid far more generously than they. As one serviceman put it, "When former officers sell their skills on the international market for profit, the entire profession loses its moral high ground with the American people."[53] Furthermore, when the executive branch of government contracts for military services, the terms of the contracts and the activities of the contractors may not be subject to congressional oversight. Employees of DynCorp working in the Balkans in the 1990s participated in sex crimes, prostitution rings, and illegal arms trading but escaped prosecution, while the company employees who blew the whistle were fired without recourse. Indigenous people will seldom recall the name of the contractor, but they will always recall that they worked for the United States.[54] Singer concludes that "PMFs are relatively free of any form of legal control to prevent or punish abuses by the firms and their employees.[55]

As the intelligence community scaled up in the aftermath of 9/11, it, too, turned to civilian contractors for data analysts, interrogators, linguists, data miners, and the like. Very often these contractors had skills, talents, education, and experience developed over years of work or study. Many had computer and electronics capabilities that were likewise much in demand. Intelligence agencies such as the NSA and the NRO often contracted with integrators such as Booz Allen Hamilton to recruit, screen, train, and manage these contractors, while government agencies oversaw the management of their security clearances—a function that was also contracted out, at least in part. The TSA, Border Security, and other homeland security agencies similarly turned to contractors to provide personnel. Often the ranks of contractors swelled until they outnumbered the government employees with whom they worked. In 2010, for example, at the peak of what political scientist Paul C. Light calls the government's "blended workforce," 7,179,000 were government employees on contract or grant, far outnumbering the 4,096,000 combined civil service, active-duty military, and postal employees.[56] This consequence of the War on Terror saw the active-duty military force increase by 1% from 2001 to 2010, while the civilian workforce in the DoD increased by 6%. The cost of contracts for services increased by 137% in the same period.[57]

In addition to the War on Terror, other forces were driving the national security community to service contracts. The overall size of government, as measured by employees, especially civil service, had been a partisan ideological issue before World War II. In general, Republican administrations in the Cold War attempted to shrink the federal work force, though not necessarily the budget, resulting in more outsourcing of jobs. Democratic administrations did the opposite. President Clinton, the last to achieve Eisenhower's goal of a balanced budget, accounted for most of the change between 1990 and 2002. In that period, the military shrank by 650,000 (about 31%) and the civil service shrank by 38,000 (3%), while contract and grant workers increased by 454,000 (24%).[58] *Washington Post* reporters Dana Priest and William Arkin discovered in their prizewinning research into secrecy in the federal government that neither they nor Secretary of Defense Robert Gates could determine how many contractors were employed by the federal government or even the Department of Defense.[59] Paul C. Light has found that the DoD workforce (military and civilian) fell by 32% between 1984 and 2017 while contractors and grantees declined by just 14% in the same period.[60] In the second quarter of 2016, the DoD had 8,730 uniformed personnel in Afghanistan and 28,626 contractors.[61]

In addition to achieving dubious savings, the use of contractors (and grantees) to thin the ranks of civil servants and military personnel also introduced problems of reliability, accountability, and loyalty in both the MIC and IIC. The largest contractor providing security and other support services to the military and other American agencies in Iraq was Blackwater USA, the brainchild of wealthy heir and former Navy Seal Eric Prince. Working with other former seals and other veterans, he formed a training and security company in 1998.[62] Blackwater achieved renown with its first major contract, rescuing Paul Bremer, the American viceroy in Iraq, from an ambush in 2003. In 2007, however, several Blackwater employees initiated a wanton firefight at Nisour Square in Baghdad, resulting in the deaths of 17 innocent Iraqi civilians.[63] This episode highlighted the lack of legal and administrative constraints on civilian contractors, in contrast with the Uniform Code of Military Justice governing American service personnel. It also raised issues of the loyalty of such contractors to the US government and its officials, as opposed to the company paying their salaries; and it begged the question of how such contractors were recruited, screened, trained, and disciplined. One official document in 2007 reported that Blackwater was billing the government $445,000 a year for a security guard. The Government Accountability Office made clear that this was not what the guard was paid nor was it really so different from the direct and indirect costs of fielding a US serviceman or woman.[64] By the following year, contractors in Iraq outnumbered US military personnel.[65] Private contractors, such as Blackwater, might have in their pay employee contractors, independent contractors, and consulting contractors, creating murky lines of responsibility, loyalty, and liability.[66] The IIC had very little legal, evidentiary, or experiential basis for designing and managing such contracts.

Nor did the intelligence community preclude contractor malfeasance. Perhaps the most damaging scandal of the war in Iraq was the mistreatment of prisoners at the Abu Ghraib prison, outside Baghdad. US military personnel, CIA agents, and contractors abused, beat, raped, tortured, and humiliated prisoners.[67] One prisoner died of mistreatment, which the military declared a homicide. Administration critics asked if any of the actions had been approved or condoned by President Bush, Secretary Rumsfeld, or other senior American officials. Eleven American military personnel were convicted of abuses and the Army major general in command of Abu Ghraib was demoted and retired. No

CIA personnel are known to have been disciplined nor was any contractor.[68] Peter Singer has raised five red flags over such military and intelligence service contracts: war profiteering, inadequate legal constraints, lack of legislative and public sanction, "a striking absence of regulation, oversight, and enforcement," and the erosion of military authority, identity, and function.[69]

These privatized military firms have become critical to an understanding of the Military-Industrial Complex for at least two reasons. First, many traditional defense contractors that have supported the MIC since early in the Cold War have expanded their portfolios from weapons and equipment into the realm of services.[70] This may be seen as a natural progression from the consolidation and diversification of the defense sector since the end of the Cold War. It may also be seen as part of a larger trend within the federal government to constrain and even reverse the growth of the federal work force. Yet another impulse, of course, has been the attempt to minimize military casualties from America's overseas interventions by limiting the number of uniformed personnel sent in harm's way. Whatever the reason, the increasing reliance on such contractors entrusts America's security to individuals whose primary loyalty may be to their corporate employers and not to the United States.

Second, sharing American security responsibilities with civilian contractors raises questions about access to the national arsenal, which has been the focus of the MIC since World War II. If, for example, the State Department hires a PMF to provide security for its foreign service personnel in dangerous environments, will it also sell to those contractors arms and equipment not available in the international arms market? And if it does, will those contractors become bona fide mercenaries, hired guns to replace active-duty military personnel? And at what level of armament—tanks, artillery, aircraft—will the line be drawn? Might the United States address its long-standing aversion to military casualties by hiring substitutes? Would the country resort to armed force more readily if its own service personnel were not at risk?

If Blackwater at Nisour Square became the most salient scandal for private security forces and Abu Ghraib for military police and interrogators, then Edward Snowden played the part for intelligence analysts. In 2013, the NSA contractor released a trove of classified documents to draw public attention to what he considered to be illegal and immoral acts by the NSA. Fleeing US prosecution and seeking asylum, Snowden ended up trapped in Russia, where he has defended his actions and called for reform of the US intelligence community in general and the NSA in particular. Among the important issues raised by his case are weak screening and oversight of contractors by the NSA and its prime contractors (in this case, Booz Allen Hamilton); high pay for contractors (the 29-year-old Snowden has claimed to have received up to $200,000 annually); misfeasance and misrepresentation by high officials in the intelligence community; illegal intrusions into communications at home and abroad by the NSA; and possible endangerment of American security interests and agents by Snowden's wholesale release of classified materials. No national consensus on Snowden's actions has yet coalesced around this exceptional case, except perhaps a widespread belief that the intelligence community and its contracting practices both warrant more public scrutiny and reform.[71]

Chapter Ten

A Peer Rival?

Reining in the War on Terror

As the federal government in general and the security community (MIC and IIC) in particular gravitated toward service contracts, government funding for innovation and development of new arms and equipment continued apace, at least for large weapon systems and similar products such as the MRAP. At the same time, both the MIC and IIC accelerated exploration of new mechanisms for government funding, especially in microelectronics and information technology, crossing boundaries between public and private realms and engaging in joint enterprises with the commercial sector. Both the military and intelligence communities are now more likely to buy COTS, more innovations arise from technology-push from the private sector as opposed to demand-pull from the government, and technology transfer involves spin-around as often as spin-off. The IIC and the MIC use similar, though not identical, contracting mechanisms while working with both their own specialized intelligence contractors and hybrid contractors who service both military and intelligence customers. As was always true of Cold War contracting with the DIB, government project managers, both military and civilian, still struggle to keep up with industry innovations in the fine print of contracts, provisions such as negotiable charges for contract changes and spare parts and perverse incentives that stretch out contracts and privilege original contractors in competition for follow-ons. The challenges of contracting with high-power legal departments of corporate conglomerates have moved Washington veteran John Alic to recommend that program management in the DoD—and by extension the intelligence community as well—be entrusted to experienced career civil servants instead of senior military officers on short rotations.[1]

Many of these issues were worked out in the dynamic environment of the War on Terror. At first, the Bush administration's military ventures in Afghanistan and Iraq went well enough. But the rationale for invading Iraq—preemption of Saddam Hussein's weapons of mass destruction—barely convinced Americans and their allies, and soon turned into another intelligence failure of historic proportions. No such weapons were ever found. After the CIA's failure to anticipate the collapse of the Soviet Union in 1991 and the 9/11 attacks in 2001, this most recent lapse precipitated the reorganization of the IC and the elevation of the new director of national intelligence over the director of central intelligence. CIA Director George Tenet, having assured President Bush that finding the weapons of mass destruction (WMD) in Iraq would be a "slam dunk," lost his job, passing through the revolving door into a portfolio of private contracts and appointments within the IIC.[2]

Public disenchantment with the war in Iraq rose through

2003 and 2004, but George Bush won re-election narrowly. The military budget continued to grow through the second Bush term, though more slowly than during the years immediately after 9/11. The base budget for the Department of Defense swelled from $287 billion in 2001 before peaking at $530 billion in the fourth year of the Obama administration (2012) and declining slightly to $523 billion in 2017. Through this period, however, the "base" was supplemented by funds for the "Global War on Terror" (2001–2009) and "Overseas Contingency Operations" thereafter (chart 4). Both of these fiscal subterfuges sidestepped the "guns and butter" stalemate that had dogged national politics ever since World War II.[3] Such exceptions to budget discipline had been invoked during the Korean and Vietnam wars, but never before had they been so large and enduring. They raised defense spending by 10% in 2001, 30% in 2010, and 15% in 2017.[4] In those same years, military spending as a percentage of GDP went from 2.9% to 4.7% to 3.1%, before starting to rise again in the administration of Donald J. Trump.[5] The rise and fall in defense spending in the George W. Bush and Barrack Obama administrations suggested no agreement on the issue of "no peer rivals," but it did suggest that there was a continuance of the Cold War consensus on maintaining a large, standing military establishment.

Perhaps the biggest change from the Cold War consensus was that the buildup and retrenchment early in the twenty-first century failed to include the kind of commitment to R&D that had been seen, for example, in the Reagan buildup of 1981 to 1987. President Bush sought some R&D increases for ballistic missile defense and Secretary Rumsfeld sought transformation,

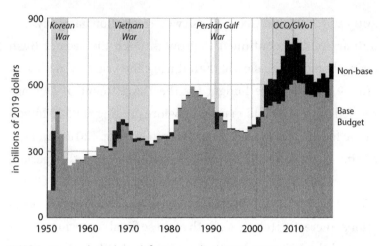

CHART 4. Base and non-base defense spending, 1950–2019. *US Congress, Congressional Budget Office,* Funding for Overseas Contingency Operations and Its Impact on Defense Spending, *CBO Report 54219, October 23, 2018, figure 1.*

not expansion, of the arsenal. The MRAP, of course, represented a concerted effort to change the arsenal on the fly. Especially in the early years of the Iraq War, air strikes were more likely to come from existing aircraft retrofitted with newer precision-guided munitions than from the next generation of planes just entering service. So, the rollercoaster of rising and falling defense budgets continued, as it had through the Cold War, with the cutbacks (generally during Democratic administrations) never quite rolling back all the increases in the previous (Republican) administrations. In general, defense spending kept rising in the first thirty years after the Cold War, as the MIC wished, in both then-year and constant dollars, even while generally heading lower as a percentage of GDP and the federal budget.

So unpopular had the war in Iraq become by 2006 that President Bush's Republican Party suffered disastrous losses in the off-year elections. He relieved Donald Rumsfeld as secretary of defense and assumed more control over military and foreign policy. In the process, he moved closer to the positions that Colin Powell and Robert Gates had been voicing for years about "no peer rivals." As early as 1991, while serving as chairman of the Joint Chiefs of Staff in the first Bush administration, Powell had questioned continuing high levels of military spending after the first Gulf War and the dawning era of the new world order. "I'm running out of demons. I'm running out of villains," he told a reporter from the *Army Times*. "I'm down to [Fidel] Castro and Kim Il-Sung," he added, referring to the aging despots of Cuba and North Korea.[6] An uncomfortable, almost impotent secretary of state in the first term of the second Bush administration, Powell had resisted the Vulcan agenda of a more muscular US foreign policy, including, at first, the decision to invade Iraq. Nor was he ever among the prominent retired senior military officers to pass through the revolving door between government service and remunerative employ in the defense industry. His roots ran deep in the military establishment, but he paid little homage to the MIC.

Joining Powell in questioning the need for ever-increasing defense spending was Robert Gates, Rumsfeld's successor as secretary of defense. Gates had served in the CIA for 23 years, with one hiatus to serve on the National Security Council, before being appointed director in 1991. After retiring in 1993, he served as secretary of defense in the last two years of the Bush administration and the first two and a half years of the Obama administration. In 2010, he shared some home truths with the Navy League, an MIC trade association group devoted to American sea power. Looking back on the War on Terror, he noted that over the previous decade the DoD had requested a nearly 90% increase in procurement, research, and development, even while presiding over a commanding preponderance of the world's naval assets. The US Navy had 57 nuclear submarines, more than the rest of the world combined. It had more combat ships equipped with the world-beating Aegis combat system than the next 20 navies combined. Its combat vessels had more total displacement than the next 13 navies combined, and 11 of those navies were allies. "Do we really need eleven carrier strike groups for another 30 years," he asked, "when no other country has more than one?" He cited the navy's failed DDG-1000 program for doubling in price while it "dwindled from 32 to seven" ships due to exorbitant unit costs. "We cannot afford to perpetuate a status quo that heaps more and more expensive technologies on fewer and fewer platforms." "Can the nation really afford," he asked his no-doubt dumbfounded audience, "a navy that relies on $3 billion to $6 billion destroyers, $7 billion submarines, and $11 billion carriers?"[7]

These rhetorical questions targeted not just the Navy League but the entire Military-Industrial Complex. Behind them one can detect Norman Augustine's Cold War quip that the costs of military aircraft were rising so swiftly that soon the services would have to share just one.[8] Also in the background lurked the ghost of Gates's immediate predecessor, Donald Rumsfeld, pressing all the services, especially the army, to stop fighting the last war, to abandon the enormous legacy weapon systems of the Cold War in favor of an arsenal tailored to the military challenges the United States currently faced. One can also hear in Gates's speech echoes of Defense Secretary William Perry's comments at the "last supper" early in the Clinton administration: consolidate now for leaner times ahead. Gates told his audience that the DoD needed "to shift investments towards systems that provide the ability to see and strike deep along the full spectrum of conflict," a phrase that stressed flexibility and innovation over legacy weapon systems designed for conventional warfare against peer rivals. He called out Congress for trying to fund programs that the military did not want, and he admitted that "the Department of Defense's track record as a steward of taxpayer dollars leaves much to be desired." The only pillar of the MIC that he chose not to blame directly was the defense industry, though it is hard to imagine that he held them blameless.

Naturally this speech reflected the views of Gates's current boss, Barack Obama, not his previous boss, George W. Bush. Obama came into office committed to ending the wars in Iraq and Afghanistan, lowering defense spending, and focusing on domestic concerns. Like Clinton before him, Obama was viewed warily by the MIC in general and the military services in particular. When a senior army officer, General Stanley McChrystal, was quoted in *Rolling Stone* magazine, along with some of his staff officers, speaking openly and disparagingly about members of the Obama administration and senior Pentagon officials, the president recalled him from Afghanistan and accepted his resignation. The episode played out in the press and attracted conspiracy theorists who argued that President Obama was "purging" the military. Presidential historian Bob Woodward wrote more temperately of the "messiness and mistrust between the White House and the military."[9] When combined with opposition in some quarters to his Affordable Care Act, Obama's alienation of hawks and conservatives contributed to Democratic loss of the Senate and House in the off-year elections of 2010. The resulting stalemate in the federal government led to the Budget Control Act (BCA) of 2011, the latest attempt to find a mechanism for balancing guns and butter.

The act sought to legislate a mandatory equivalence between military and nonmilitary discretionary spending—between guns and butter—while at the same time trying to encourage reduction of the federal budget deficit. It imposed tighter discretionary spending limits, and it prescribed automatic, mandatory spending reductions each year that Congress failed to reach a deficit reduction agreement. The overall goal was to achieve an average savings of $109 billion per year over nine years and thus reduce the growth of the national debt by $1.2 trillion.[10] The spending reductions were often referred to as sequesters. Congress might appropriate more, but some percentage of those appropriations would be sequestered if the bill's criteria were not met. Congress was essentially tying its own hands as a way of forcing itself to resist the perennial temptation of deficit spending. It did, however, allow two exceptions—for "emergency requirements or for overseas contingency operations."[11] These overseas contingency operations, or OCOs as they came to be called, proved to be a loophole big enough to drive an MIC through. The OCO essentially replaced the Bush administration's "Global War on Terror" as an escape clause from the Budget Control Act.

The BCA, scheduled to expire in 2021, did have some positive impact. In general, however, Congress has acted consistently

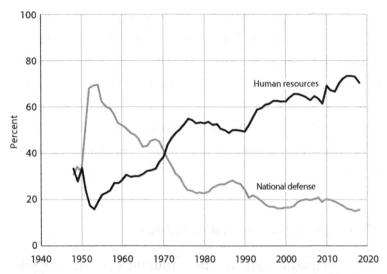

CHART 5. US defense and human resources spending as percentage of federal budget, 1948–2018. *United States, White House, Office of Management and Budget (OMB),* Historical Tables, *table 3.1.*

to blunt the impact of the BCA on both defense and nondefense spending.[12] As previously noted, since 2001, "non-base funding," that is, funding for the "Global War on Terror" or OCO, had averaged about 20% of the DoD's budget (chart 4). In other words, it added about 25% a year to the base budget. Furthermore, from 2006 to 2018, about $50 billion per year (in 2019 dollars) on average of OCO funding had really been diverted to "costs of enduring activities rather than the temporary costs of overseas operations."[13] In other words, the DoD had used the OCO exemption in part to get around the limits imposed by the BCA.

In the early years of the Trump administration, defense spending predictably began rising again. Most importantly, both Republicans and Democrats disregarded annual budget deficits, which had declined for a while in the Obama administration, and the cumulative federal deficit, which accelerated in the face of a major tax cut and increased spending for both guns and butter. This growth in both the DoD base budget and the OCO, however, reversed but slightly the general downward trend of defense spending as a percentage of GDP. Since the end of the Cold War, that number fluctuated between 2.8% and 4.7%, lower than at any time since the brief demobilization following World War II.

The Changing Innovation Paradigm

Other comparisons and contrasts with the Cold War era MIC mark the security landscape of the first two decades of the twentieth century. There has been, for example, a blossoming of public-private partnerships investing in commercial enterprises that also promote military innovation. Linda Weiss, in *America Inc.?*, identified some of these ventures as part of a National Security State (NSS) because they involve multiple agencies, not simply the DoD or the IC. For example, the DoD, the National Geospatial-Intelligence Agency, and the Army's Communications Electronics Command jointly created Rosettex, a venture capital fund organized under contract with a legacy research agency, SRI International, to serve the National Technology Alliance (NTA), which had been formed in 1987 to promote "commercial and dual-use technologies."[14] Federally funded research and development centers (FFRDCs), an institutional form going back to World War II, added to their ranks the Venture Acceleration Fund in 2015, a start-up formed to manage the Los Alamos National Laboratory for the Department of Energy. It now focuses on funding economic growth in northern New Mexico under the direction of a regional development corporation.[15] Partnerships such as the MIT-based Institute for Soldier Nanotechnologies "mobilize the resources of industry, academe, and government in dedicated research and engineering centers and industrial consortia."[16]

None of the kaleidoscopic array of public-private collaborations that Linda Weiss chronicles are entirely unprecedented.[17] Many go back to World War II and even earlier. What has changed since 9/11 is the relative decline in the traditional mechanisms of government promotion of research and development and the unshackling of government agencies to pursue new collaborative mechanisms with the private sector. Weiss's NSS still contracts for R&D from universities, industrial and independent laboratories, and manufacturers of its weapons and equipment, but this type of funding always had something of a top-down, demand-pull flavor to it. The government specified what it wanted, and institutions were contracted to develop and produce it. Since 9/11, however, more and more technology useful to the military and the IC bubbles up from the commercial sector, often from start-up companies built around a new product or capability, in a process of technology-push. Agencies of the NSS need to be attuned to the marketplace, especially in microelectronics and information technology, and nimble in discerning and nurturing the nascent technologies that might enhance their capabilities. Private and commercial sectors now account for 98% of demand for semiconductors. The military's 2% exerts little impact on the market and raises serious concerns about guaranteeing access to critical components.[18]

What this study calls the microelectronics domain corresponds more closely now to C⁴ISR—the acronym for command, control, communications, computers, intelligence, surveillance, and reconnaissance.[19] The common denominator of these military functions is information, the horizontal sharing of knowledge and guidance sought in net-centric warfare, at the heart of the Revolution in Military Affairs. All these capabilities have taken their current form and power from the microelectronics domain.

For the military establishment to realize the maximum gain from the rapid and complementary advances being made in the technologies behind C⁴ISR, it must be alive to the possibility of spin-around, commercial technological innovations that can be adapted to military purposes, and military or intelligence developments that can morph into commercial products and services. Night vision goggles, for example, might have been developed for military purposes, but if hunters and other civilians purchase commercial models on a large scale, unit costs will come down in both public and private sectors. Furthermore, to the extent that military functions can be served by commercial models, the government can avoid the gold-plating that so often went into technologies custom-designed for the military.[20]

Of course, the old risk of "national industrial policy" remains, with obvious allusions to socialist or communist societies, but concerns on this front appear to have subsided for three reasons. First, military and intelligence spending by the federal government accounts for less than 5% of GDP, sometimes less than 3%, hardly the 10% or more sometimes reached during the Cold War, let alone the 41% of World War II. Second, most of the public and private collaborations into which the military and intelligence communities have inserted themselves have been comparatively small and transparent, open to potential competitors and resistant to monopolization. Third, these collaborators have usually targeted comparatively new companies and innovations, seeking to nurture them early in their development, not to privilege a mature company or technology over its competitors. Indeed, this diversity of funding mechanisms has been shown to be one of the hallmarks of modern American innovation in general.[21] Virtually all instances of US government promotion of technological innovation pale in comparison to the market disruptions caused when the government intervenes to bail out a major contractor, as was often done in the Cold War, in order to maintain critical manufacturing capabilities. In 1971, for example, as Lockheed Aircraft Corporation struggled with parallel financial crises in its C-5A military transport and its commercial L-1011 passenger jet (following on the heels of a bribery scandal over foreign sales of its disappointing F-104 interceptor), the US government bailed it out with a $250 million loan guarantee and other favors that nursed it back to profitability.[22] Then Lockheed rose to become the nation's—often the world's—largest defense contractor, a position it held for most of the first two decades of the twenty-first century. Along the way, it absorbed some competitors from the 1960s and 1970s, such as Martin Marietta, while witnessing the demise or assimilation of others, such as McDonnell Douglas and Rockwell, both of which were acquired by Boeing, the perennial number two defense contractor. The old system of oligopoly, coddling a small number of powerful collaborators, distorts market forces far more than the venture innovation of twenty-first century collaborations.

Thus, two separate boundaries are being blurred by the evolving military-industrial relationship. The challenge to the civil-military boundary that ensures civilian control of the military was more apparent than real, because civilians in the Department of Defense controlled virtually all these joint ventures. But the public-private divide protecting the country—in the minds of some observers—from national industrial policy was certainly eroding. This latter pattern fortified the claims of Michael Hogan, Linda Weiss, Anne Markusen, and other scholars who believe that the United States has become a "National Security State."[23] Indeed, Weiss believes that the transformation of military and intelligence relations with industry since 9/11 has gone so far beyond the excesses of the Cold War that "President Eisenhower's warning against the Military-Industrial Complex is not simply outdated, but has been firmly laid to rest."[24]

Recent developments have also resurrected the Cold War dispute over the impact of military and other security spending on the national economy. For example, DARPA, the advanced-research arm of the DoD, has long shared President Eisenhower's view that a strong American economy goes hand in hand with strong national security. DARPA's critical contributions to computer development in the United States were informed by the goal of creating a sound technological base for military applications.[25] But the contest over guns and butter leading up to the Budget Control Act of 2011 revealed that many, if not most, American political leaders saw discretionary spending as a zero-sum competition between national security and social programs. The failure of the BCA to quell that perception reflects its continuing relevance, even though many have pointed out the oversimplification inherent in such a binary. Robert Higgs, for example, has offered a libertarian analysis suggesting that the opportunity costs of defense spending fall not only on the nondefense, discretionary spending of the federal government but also on "all private purchases, whether for consumption or investment." Using these three variables, instead of the conventional two, he concludes that the private sector compensates for virtually all the opportunity costs imposed by defense spending.[26] It appears that the "guns and butter" metaphor captures a salient notion in American public consciousness without necessarily revealing the true complexity of the economic impact of military spending.

Perhaps no realm of military innovation and acquisition captures the tenor of the three decades since the end of the Cold War more vividly than the air force's pursuit of a fifth-generation fighter aircraft. Determined to maintain its Cold War air superiority over the battlefield, the air force pursued stealthy fighter aircraft, whose radar could "see" and fire upon enemy planes before being seen and engaged themselves. Having pioneered stealth technology, which proved difficult to replicate, the air force sought to maintain that edge, even at the expense of other performance characteristics.

Before the Cold War ended, the air force began developing an "Advanced Tactical Fighter," that is a fifth-generation fighter with stealth technology. To meet the anticipated Soviet threat, the air force planned to acquire 750 of these planes. When the air force selected the Lockheed Martin F-22 Raptor in 1991, the collapse of the Soviet Union and the continuing decline in US defense spending were already eroding the air force's need for such an airplane. Rising development and production costs and further defense cutbacks through the decade of the new world order led to cost caps on the F-22 program, which could be met only by cutting the number of planes on order and thereby raising the unit cost of each aircraft. In the end, only 187 F-22s were acquired at a total program cost of $350.5 million per plane, compared with the navy's concurrent acquisition of 493 F/A-18 Hornets at $93.9 million per jet.[27]

Disappointment with the F-22 was overdetermined. The air force had anticipated a Soviet threat that never materialized. Lockheed Martin bid low on a contract that experienced predictable delays, cost overruns, and underperformance. The air force added expensive capabilities as development and production proceeded. Defense spending was contracting as F-22 costs rose, forcing a shorter production run and driving up unit costs. Further, the F-22 proved to have high operating costs and little demand for its unique capabilities in the wars in Iraq and Afghanistan. In the end, this fifth-generation fighter saw little combat. As Air Force Lieutenant Colonel Christopher Niemi reported, "Since becoming operational, the F-22 has conducted only deterrence deployments and homeland defense intercepts—missions hardly worthy of its unmatched prowess and cost."[28] One critic of the MIC has called it "the fighter without a foe."[29]

In spite of a resolve to never again commit the mistakes of the F-22 program, the air force in 1996 selected the same contractor, Lockheed Martin, to develop and build a follow-on fifth-generation fighter for which there was still no foe.[30] The F-35 "Lightning II" Joint Strike Fighter, which is not yet fully operational a quarter-century later, proceeded to repeat all the mistakes of the F-22 program and added some new ones. It insisted on stealth technology. It prescribed capabilities that were not yet developed and demonstrated. It predicted an optimistic delivery schedule, hoping to meet it by concurrency, the risky tactic of beginning production even while still developing some components. If those developments do not turn out as predicted, expensive retrofits must be applied to all the airplanes already begun. As one reporter put it, "The plane was built as it was being designed.[31]

Making the challenge still greater, the DoD agreed to produce three different versions of the plane with 80% commonality of major components, in spite of the fact that the F-111 and other joint development programs from the Cold War had never achieved their commonality targets.[32] The air force version would be a fighter/interceptor maintaining air superiority over the battle space. The navy version would take off from, and land on, carriers. A marine corps version would be able to take off from short runways and land vertically, allowing for retirement of the troubled Osprey. Of course, having three service sponsors expanded the military-industrial-congressional base of support for the program. To further increase orders for the plane and thereby hold down unit costs, the F-35 program also included an unprecedented number of foreign partners, nations that would contribute to development and production and buy the aircraft when it became operational. The United States had engaged in many coproduction agreements during the Cold War, especially with its NATO allies,[33] but the F-35 set new standards for cost- and technology-sharing and for sales of state-of-the-art technology. Eight "partners" each agreed to contribute between $100,000 and $2.2 billion to the program, while four other participants agreed to cooperate and buy aircraft.[34] So ambitious was the program and so damaging the experience of the F-22 that it became a cliché to observe that the air force had "bet the farm" on the F-35.[35]

Two decades later, the farm is still at risk.[36] As of 2015, the program's costs had more than doubled and it was at least seven years behind schedule.[37] Many of the specified capabilities had been simply dropped, and operational constraints were placed upon existing planes because of fuselage cracks and other unanticipated problems. Lockheed's performance as prime contractor was no better, and perhaps worse, than it had been on the F-22. When Air Force Lieutenant General Michael Bogdan took over management of the floundering program in 2012, he reported that "it was the worst relationship [between the government and a contractor] I had seen in my acquisition career. I had a sense, after my first 90 days, that the government was not in charge of the program."[38] He put Lockheed on notice that the air force would no longer pay for "mistake after mistake after mistake."[39] Todd Harrison, an aerospace expert with the Center for Strategic and International Studies, observed that the commonality goals for the program were unrealistic from the start, because combining requirements for the three services was bound to produce "an aircraft that is suboptimal for what each of the services really want." In the case of the F-35, he added, "The three versions of the plane actually don't have that much in common."[40] Michael Hughes reported in 2015 that the targeted 80% commonality of the three versions had fallen to between 27% and 43%.[41]

In spite of such problems, however, the program survived, like "the born-again bomber" and the last Seawolf submarine.

Even so powerful a critic as Senator John McCain, who in 2016 called the F-35 program a "scandal and a tragedy with respect to cost, schedule and performance," recognized that it was immune to cancellation. With more than 1,500 subcontractors, 133,000 jobs spread out over 45 states, and a 48-member Joint Strike Fighter Caucus on Capitol Hill, the program enjoyed "basically a veto-proof constituency," according to Dan Grazier of the Project on Government Oversight.[42] But, at the time of this writing, it is impossible to predict if it will ever be considered a success. The price of the F-35A, the air force model, has fallen from $241.2 million per plane in 2006 to $89.2 million in 2018. But in 2017 and 2018 only half the fleet was operable on any given day and maintenance costs were projected to account for more than 70% of total costs over the life of the program. In 2018, the F-35A cost about $40,000 an hour to fly, roughly double the cost of the navy's F/A-18E/F.[43] And some international partners in the program are retreating from earlier commitments to buy significant numbers, raising the unit costs for the other buyers.[44] In 2019, the United States removed Turkey, a NATO ally, from the F-35 program, canceling its order for 100 of the planes and removing from the supply chain 900 parts that were provided by Turkey. The action followed Turkey's acquisition of a Russian missile defense system, raising concerns that the deal would give Russians access to American stealth technology in the F-35.[45] The final cost of the program—estimated at $1.5 trillion for about 2,443 planes—would come out to about $614 million per plane, but it is far too soon to know the total costs or the number of planes that will be bought.

What, then, is to be made of the F-35? It is certainly the

world's most advanced fighter-interceptor, boasting stealth technology that Russia and China have yet to match. Though many of the plane's promised capabilities have failed to perform as planned, it incorporates such state-of-the-art features as advanced sensor-weapons integration and helmet display giving the pilot 360 degrees of vision. Its user-friendly controls have been a hit with pilots, who liken flying it to playing video games—Air Force Chief of Staff David Goldfine called it "a computer that happens to fly"—and praise its responsiveness and agility.[46] Its systems are so fully automated that the lone pilot can manage all its functions without copilot, navigator, or weapons officer.

Its shortcomings are equally remarkable. Like the F-22, its stealth coating requires constant maintenance. The stealth design also limits its speed, range, weaponry, and maneuverability. It is slower than the most recent Russian and Chinese designs, and while it can engage those planes before being detected and attacked, it may not fare so well in dogfights at closer ranges. Neither the F-22 nor the F-35 has ever engaged in aerial combat, and until that happens it will remain impossible to know if they justified their cost. The Chinese J-31 fighter, a knockoff of the F-35 based on hacking of American computers, suggests that the American military policy of staying a generation ahead of its adversaries may be increasingly difficult to sustain. One informed critic has called the F-35 "an inherently terrible airplane."[47] Lockheed's general manager for the program from 2000 to 2013 has said that "the technology to bring [all the F-35's capabilities] together in a single platform was beyond the reach of industry."[48]

The fate of the F-35 hangs in the balance as this study goes to press. Its story—and its uncertain future—reflect the current state of the Military-Industrial Complex three decades after the end of the Cold War. The story pivots in 2001, at the transition between the new world order and the ensuing War on Terror. As the F-35 program formally began, the attacks of 9/11 redefined the military threat facing the United States, and Donald Rumsfeld entered office as secretary of defense determined to "transform" the American military into a post–Cold War institution tailored to America's security needs in the twenty-first century. Rumsfeld oversaw the cancellation of the army's Cheyenne helicopter and began the critical review that led his successor, Robert Gates, to limit production of the F-22. Rumsfeld did not, however, cancel the F-35 program, which Australian aviation expert Carlo Kopp was just then calling a "Cold War Anachronism." Kopp's prescient insight was that "whether one is hunting a high technology Russian mobile SAM system, a mobile ballistic missile system, or a bunch of terrorists in a four wheel [*sic*] drive ... the best technique is loitering bombardment which is not the forte of smaller fighters—including the Joint Strike Fighter." [49] The F-35 embodied the same kind of thinking that led Rumsfeld to cancel the Army's Crusader self-propelled howitzer. It was a weapon system designed for the last war and ill-suited to the wars the United States was about to enter.

But how could a political appointee such as Rumsfeld overrule the industry and military experts—to say nothing of their congressional supporters—on technical judgments about complex weapon systems? When aviation expert Loren Thomson consulted with Rumsfeld early in the Bush administration on the Cheyenne helicopter and F-22 fighter, he discerned that the defense secretary "knew next to nothing about either program"—

except their "astronomical price tags."[50] Nor could Rumsfeld have known that the F-35, just then getting started, would turn out to be even more problematic. More than a decade after the Cold War, the MIC continued to drive spending ever upward in pursuit of cutting-edge innovation while the responsible officials in the executive branch tried to hold spending down. It was the same tug of war that worried Dwight Eisenhower half a century earlier.

In the decade and a half since the first flight of the F-35, three other emerging technologies have proven more critical to US national security: unmanned aerial vehicles (UAVs, popularly known as drones), cyber technologies, and robotics.[51] Not all of them combined have consumed the funding invested in the F-35, but all have proven themselves in combat and all are on the front lines of America's national security.

UAVs have become the instrument of choice for striking nonstate actors bent on terrorism against the United States and its interests.[52] The vehicles themselves evolved from reconnaissance aircraft, in much the same way that helicopter gunships evolved from helicopters deployed for transportation of troops, supplies, and casualties; both aircraft types were dual-use technologies. Though neither weapons platform is truly autonomous—drones are piloted remotely—both may be seen by some as a result of technological determinism—in this case to arm platforms that can position weaponry advantageously. Armed UAVs also provide the United States—as well as Israel, the country that pioneered them—a safe, inexpensive stand-off weapon providing surveillance and lethality to fight nonstate actors in operations other than war. The legal, ethical, and even pragmatic arguments raised against them have failed to overcome their appeal to even so humane a president as Barack Obama, though American use has declined precipitously since its peak in 2010.[53] Furthermore, the weaponization of UAVs is still in its infancy. Michael J. Boyle has identified some of the worrisome directions in which this technology might evolve, especially autonomous drones, swarming of small drones, and miniaturization.[54]

Cyber, another dual-use technology from the microelectronics domain, remains at present an emerging threat not yet fully weaponized. Indeed, some experts insist it is not a weapon and will not become one.[55] The DoD disagrees, calling cyber the "fifth domain" of warfare, along with land, sea, air, and space.[56] In 2009, President Obama allowed creation of a US Cyber Command to work with the NSA. President Trump elevated it in 2018 to become one of the DoD's 11 combatant commands. Of the roughly $13 billion spent on cyber by US federal agencies in 2017, the DoD and Homeland Security between them accounted for two-thirds.[57] The private sector invests a much larger amount in cybersecurity, most of it defensive but some offensive as well. Shared concerns about cyberattacks for espionage, disruption, and even physical damage demand government cooperation with private businesses and institutions, further blurring the lines of public-private and civil-military relations. Though the United States has been publicly restrained in its responses to hostile cyberattacks by state and nonstate actors, Shane Harris believes it has a reputation as being fairly aggressive in secret. Harris further believes that the US private sector will play the major role in defending "critical facilities" in the United States.[58]

The third area of emerging technologies is robotics. Of

course, drones and even cyber "worms" can be seen as robots of a sort, because they highlight one of the main concerns about all such weapons—their autonomy. All three can be more or less autonomous, depending on how they are designed and operated. One quality of the artificial intelligence that will govern so much of their behavior in the future is learning. It is important that their programs allow them to adapt to their environment and learn from experience. But how do you constrain that learning so they stop short of becoming HAL, the computer from the 1968 science fiction movie *2001: A Space Odyssey*, that learned how to take control of his human masters?

P. W. Singer believes "war-bots" of the future will perform difficult and dangerous military missions while keeping humans out of harm's way.[59] They will operate on land, at sea, and in the atmosphere and space. And they will vary in size from warships to nanotechnology. The key issue is whether they will be under continuous human control or set loose to conduct prescribed missions autonomously. If the latter, will they be programmed to kill? If so, will they cross some ineffable threshold into a world no one will want to inhabit?[60]

The MIC continues to refine these and other military technologies of the future, with unpredictable consequences. Some analysts attempting to discern the future trajectory of American strategy and R&D see possibilities ranging from a new RMA focused on swarming microbots to a third, offset strategy to operationalize technologies of autonomy and artificial intelligence within a strategic doctrine.[61] All such predictions, however, turn on the great unknown: what strategic missions the US military will be called upon to perform in the years to come. It seems likely that the armed forces will move slowly, perhaps reluctantly away from the great legacy platforms of the Cold War and toward ever greater reliance on the lean, ingenious, adaptive technologies of the microelectronics domain.

Conclusion

T he Military-Industrial Complex abides. It should really be called the Military-Industrial-Congressional Complex, but that was true even in Eisenhower's time. This complex still consists of *a convergence of state and industrial forces collaborating to shape US national security policy to privilege the forces' respective special interests over the national interest and national security.* All parties to the complex seek increased military spending, which remains the best evidence of their existence and barometer of their impact. The continuing presence and influence of the MIC depends upon the survival of a large, standing military establishment, the absence of a great-power war since World War II— what historian John Lewis Gaddis calls "the long peace"—and a shared faith in the quality of American arms, not the size of American armed forces.[1] Yet this continuing pattern does not necessarily support the conclusion that the United States has become a National Security State. Indeed, military spending consumes a significantly smaller share of national income and federal spending than it did during the Cold War.

While most literature on the MIC treats the term pejoratively —as did Eisenhower when he introduced it—this study has attempted to remain neutral on the institution's net impact.[2] The MIC has safeguarded Americans and their interests in a dangerous world, while simultaneously ignoring and eroding many of America's most fundamental legal and political principles. Like many human institutions, it has proven to be an effective but imperfect instrument of the national will.

The categories of analysis that organized the discussion of the Cold War in part I display more continuity than change since 1990. *Civil-military relations* have undergone a blurring of distinctions between the two realms, though the fundamental principle of civilian control of the military remains strong. In the exalted tradition of George Washington, George C. Marshall, and Dwight Eisenhower, Colin Powell moved from the highest military ranks as chairman of the Joint Chiefs of Staff under President Bill Clinton to national security advisor for President Ronald Reagan and secretary of state under President George W. Bush (figure 10).[3] Powell is, in many ways, the iconic citizen-soldier of the post–Cold War United States, combining distinguished military credentials and combat service with astute political and diplomatic skills for navigating both Democratic and Republican presidencies and congresses. Among his many contributions to civil-military amity, he became the first chairman of the Joint Chiefs of Staff to serve his entire term under the reforms of the Goldwater-Nichols Act (1986), which greatly increased the power and authority of the chairman. He complemented his civilian boss in articulating the Weinberger-Powell Doctrine on foreign deployments. In the process, he contributed significantly to the recovery of the military's reputation after the Vietnam War. This recovery, now deeply rooted in American

FIGURE 10. Secretary of State Colin Powell holds up a vial of anthrax to illustrate to the United Nations Security Council the threat posed by Saddam Hussein's alleged program of weapons of mass destruction. Powell's testimony, based on faulty intelligence, convinced many in America and abroad to support the US Coalition to invade Iraq in 2003. Powell later said his testimony would always be a "blot" on his career. *Mark Garten, United Nations photo #24771, February 5, 2003.*

consciousness, occurred despite the steady decline of veterans in government service, especially in elected office.

The Goldwater-Nichols Act also created unified commands that distributed important assignments more evenly among the services and it restrained interservice rivalry for roles and missions, though in procurement it led to the kind of mutual back-scratching that produced the F-35.[4] Unlike the precursor F-111, forced on the services by Secretary of Defense McNamara, this similarly futile pursuit of cost savings through commonality was conceived by the services themselves as a way of building support for the air force's attempt to do over the F-22.

Powell's example of studious nonpartisanship was undermined by a disturbing trend of some retired officers exploiting the mantle of armed service to support partisan candidates and causes. The sad example of Medal of Honor winner Admiral

James Stockdale running for vice president in 1992 on the independent ticket of Ross Perot diminished the candidate, the navy, and the armed services in general. The muted sacrifices of James Mattis, H. R. McMaster, John E. Kelley, and others to restrain the most imprudent impulses of the inexperienced President Donald J. Trump stand in stark, if perhaps futile, contrast.[5]

There has also been some blurring of the Cold War boundaries between *state and industry* since the Cold War, more retired officers have served as executives and board members for defense and intelligence contractors, and some have passed once or more through the revolving door separating the public and private sectors. On a more positive note, the government expanded its principal-client relationship with the private sector by engaging in joint ventures and "partnerships" with industry. Furthermore, new companies flourished as integrating contractors for national security agencies, offering both products and services. At the outer limits of these new realms of state-industry relationships were service contracts for security and training that often bordered on the realm of the mercenary. This pattern reflected in part the public's aversion to casualties among service personnel and a simultaneous indifference to contractor deaths, costs, and misfeasance.

The greatest changes in relations *among government agencies*, besides Goldwater Nichols, occurred in the aftermath of 9/11. Government-wide reorganization included creation of the Department of Homeland Security, which represented the most profound transformation of national security since World War II. The Department of Defense (DoD), created in 1947 and modified in 1949, was really an institution of national force projection. Even its most defensive function, deterrence of strategic attack with nuclear weapons, was based on threatened retaliation —i.e., counter attack—as opposed to physical defense. President Reagan's Strategic Defense Initiative has never achieved the capabilities he sought, nor have subsequent attempts at ballistic missile defense succeeded beyond the tactical and theater levels. President George W. Bush captured the more fundamental ethos of American national security in the Cold War and beyond when he announced, following 9/11, that the United States would defend itself against future terrorist attacks by seeking out the terrorists around the world, preempting or preventing their attacks. The best defense, he proclaimed, in the old military chestnut, is a good offense; and it has always been offense to which the US DoD directed its efforts and resources.

But the creation of the Department of Homeland Security, the reorganization of the intelligence community, and other structural and organizational changes made after 9/11 actually focused significant parts of the American security establishment on defense. These new institutions have been more effective than the DoD in carrying out their assignments. Tasked essentially with nation-building in its two, post–9/11 wars, Iraq and Afghanistan, the DoD has struggled to complete a mission for which it was not designed, equipped, or trained. American servicemen and women have generally performed heroically and competently, at great cost in physical and psychological harm, without achieving security and self-government for their host states. As Benjamin Fordham has observed, "The irony of American military supremacy is that it makes the nation more likely to find itself involved in the unconventional wars for which its capital-intensive military force is least well-suited."[6] Meanwhile, the United States has proved generally impervious to continued attempts by terrorist organizations around the world to repeat anything like 9/11 in the United States. This achievement is a triumph of intelligence-gathering and coordination, screening of people trying to enter the US, and identification and tracking of people who do enter the US with malicious intent. The US has proven more vulnerable to domestic than international terrorism since 9/11.

The relationship between the national security community and the *scientific-technical community* suffered more from the Vietnam War than from any other occurrence since World War II. Vannevar Bush's World War II model for harnessing science and technology in service of the state never came to fruition. But early in the Cold War, it nonetheless achieved the kind of military-civilian cooperation he advocated, in the services themselves and in a research infrastructure woven into the public and private fabric of US R&D. But opposition to the Vietnam War on US campuses soured the tradition of cooperation that had flourished in World War II and led to institutional and ideological barriers to military work on America's campuses and beyond. Such animus abides in the resistance, for example, of workers at Google to cooperate with military and intelligence agencies, a reluctance fueled by the revelations of abuses within the military and intelligence community made by Edward Snowden and other critics of the National Security State.

Vannevar Bush also illustrates a less salient but nonetheless profound characteristic of government relations with the scientific and technical community since World War II. Bush was part of the legion of scientists and engineers who produced Franklin Roosevelt's arsenal of democracy in World War II, making it the first war in world history in which the arms in play at the end of the war differed significantly from those at the beginning. It was this pattern of innovation that Bush sought to institutionalize with *Science: The Endless Frontier.* Eisenhower took up the baton after the technological shock of Sputnik in 1957, creating NASA to meet the Russian space challenge and the Advanced Research Projects Agency to promote military innovation, proposing the National Defense Education Act to fund the science and engineering students of the future, and empaneling a President's Science Advisory Committee (SAC) to ensure that the president had his own scientific and technical advisors on call.

Alas, this example has atrophied in the years since Eisenhower called out the Military-Industrial Complex. His warning of a scientific-technical elite offended and alienated many scientists and engineers, including some who served on his own SAC, though Eisenhower tried to mend that fence. But other presidents, such as Richard Nixon, Ronald Reagan, and Donald Trump made no secret of their disdain for scientists who espoused policies that were not conducive to national security and economic growth—for example, arms control, environmental protection, climate change, and so on.

Throughout the Cold War and beyond, the promotion of new military technologies has been shaped in part by the relationship between *society and technology.* Growing public concerns about technological determinism and autonomous technology have often manifest themselves in alarm over nuclear, biological, and chemical weapons, mutual assured destruction, napalm, agent orange, land mines, arms transfers, intrusive surveillance, drone warfare, and the transfer of the "baroque arsenal" to unstable or rogue states and nonstate actors. When an authority as sane and astute as C. P. Snow could warn that nuclear war was inescapable, average citizens could be forgiven for concluding that the modern technological arms race would end in Armageddon, under the misguided direction of our own *Dr. Strangelove* or the machinations of some other evil genius. The end of the Cold War without the euphemistic "exchange of nuclear weapons" so often predicted by the technological determinists helped to quiet such public alarms, but the failure of President George H. W. Bush's "new world order" and the threat of massive terrorist attacks with box cutters and IEDs stirred new concerns that unbridled technological innovation could yet threaten the liberal world order. The demonstrated power of cyber warfare only contributed to the suspicion that even a development as positive, universal, and democratic as the internet could be weaponized by enemies of limited means to inflict massive casualties and damage on the most developed states, whose reliance on networked infrastructure made them the most vulnerable to such attacks.

Finally, the *international arms trade* also displayed some of the autonomy and ambiguity that characterize technology in general. In spite of its announced intention to limit arms transfers after the Cold War, the United States displaced the Soviet Union as the world's leading arms exporter and accelerated, with the F-35, the process of sharing its most advanced military technology. The temptation to reduce unit costs of the F-35 drove the United States to sell the airplane in its prime and hasten the arrival of its technological obsolescence, thereby assuring that another generation of replacement aircraft would have to begin development even before production of the F-35 had run its course. Furthermore, Chinese hacking of the US defense industrial base (DIB) had already ensured that a knockoff of the F-35 was in production and ready for sales to those states that did not want to pay the US price. Meanwhile, low-level intrastate and interstate wars went on around the world with a seemingly limitless supply of arms and ammunition, some of it sold in Faustian bargains with arms dealers who take out mortgages on the futures and assets of the states, combatants, and parties with whom they contract for their goods and services. There are, in short, multiple agents in the international marketplace willing and able to deliver on credit products from the arms bazaar in return for future claims against the assets and even the political control of the combatants they service.

Part II of this study has narrated the evolution of the MIC in the past 30 years, asking if it looks more or less as it did during the Cold War. In short, what changed and what remained the same?

The three essential preconditions for the complex remain unchanged: The United States continues to rely upon the quality of its arms and equipment over the quantity of its armed forces; it continues to maintain a large, standing military establishment in peacetime; and there has been no great-power war since World War II. As long as those three conditions continue, it is difficult to see how the MIC might disappear. It is, as historian Mark Wilson has observed, "an enduring and resilient institution."[7]

All of the old transgressions of the MIC revealed in part I of this study rear their sturdy heads in part II. Buying-in or lowballing or Ron Smith's "conspiracy of optimism" are so shamelessly pursued and accepted that they are taken as normal and predictable business practice in the MIC and the IIC. They are even imitated in other government agencies.[8] This produces such astonishing results as the leading defense contractor, Lockheed Martin, getting the nod to develop and produce the F-22, which proved so costly in the end that its production run was cut back by 75%. Then, when the air force felt obliged to build yet another fifth-generation fighter, because the F-22 was too expensive to buy and fly, it chose Lockheed Martin again. This encore led to cost overruns in the world's most expensive weapons program ever: the F-35. Some characteristics of the MIC appear impervious to reform.

Corporate welfare and bailouts also continue, accelerated by the consolidation of the aerospace industry, which was encouraged and abetted by the government at the end of the Cold War. As the number of prime defense contractors shrank and the size of the integrators swelled, more and more became too big to fail. Lockheed and Lockheed Martin have been bailed out several times. Newport News Shipbuilding remains the only contractor for aircraft carriers, immune from both competition and collapse. Individual projects have been canceled: the Crusader, DDG-1000, and the Cheyenne, for example.[9] But the weight of sunk costs can lead otherwise prudent managers to throw good money after bad. The 2019 GAO annual report to Congress evaluated the $1.69 *trillion* portfolio of 82 major weapon systems acquisitions. It noted that the average age of these programs was 14.2 years and growing. This results in part from introducing "new capabilities and upgrades through additions to existing programs"—a kind of gold-plating that raises costs by evading more rigorous evaluation.[10]

The contract remains the instrument of choice for acquisition of goods and services, especially for the large, high-tech weapon systems that take pride of place in the US arsenal. Government efforts to limit contractor abuses have failed to change the Cold War balance of power between principal and agent. Contractors still have perverse incentives to stretch out development on cost-plus contracts. Exempting spare parts and auxiliary equipment from the cost controls imposed on major systems continues to permit wretched gouging of the government. Contract provisions that allow companies to recover losses on one phase of development by including them as costs in the follow-on phase encourage sloppy performance, lax accounting, and outright fraud. Cancellation charges favoring contractors can be so generous as to render the contractor virtually harm-free, as was the case with the planned cancellation of the third Seawolf. Sole-source contracting, a long-standing abuse, is facilitated by the very consolidation of defense contractors that the government promoted. The GAO reported in 2019 that 67% of "82 major weapon systems acquisition programs" were "not competed."[11]

The acceleration of government contracting after 9/11 increased the opportunities for waste, fraud, and abuse even while it undermined the government's ability to monitor contracts and account for spending. As Paul Light's *The Government-Industrial Complex* has demonstrated, the US government cannot even identify all its contractors, let alone account for all the funds paid out to them.[12] Economists Mark Skidmore and Catherine Austin Fitts have discovered that the dollar amount of unidentified funds being carried on the DoD's books swelled from $2.3 *trillion* in 2001 to $6.5 *trillion* in 2015.[13] The largest slice of government contracting goes to the military, which also has the poorest record of accounting.[14] The great increase in contracting for services following 9/11 only made a bad problem worse.[15] The legal and administrative rules governing contracting for services are less refined than those for development and acquisition of material goods, and contractors become government agents for whom the rules and responsibilities are rudimentary. All of these contracting weaknesses erode the principal-agent relationship and diminish the likelihood that American policies are being pursued as intended.

The revolving door between the public and private sectors continues to turn, with the same ambiguous effect. In the abstract, it makes sense for experienced military personnel, many of whom are forced into retirement in their forties or fifties, to find employment with defense contractors, where their years of experience can be put to good use. And it makes equal sense for employees of defense contractors to fill occasional posts in the civil service, where they can help bridge the gap between public and private sectors and military and civilian practice. There is no reason to believe that such transplants are necessarily corrupt.

In practice, however, the opportunities for mischief have often overwhelmed the better angels of these migrants, almost always to the advantage of the contractors and the disadvantage of the government. Too often the special interests of individuals and the companies that employ them seem to overwhelm the national interest. Sometimes such malfeasance arises from an inability to recognize or appreciate a conflict of interest. Other times it arises from an unabashed, even willful, insistence that the special interest and the national interest are identical. What is good for General Motors is good for America. Some veterans who have passed through the revolving door more than once—Mike McConnell comes to mind—have convinced the powers that be on both sides of the public-private divide that they can serve both masters honorably and effectively. Some who have risen in industry have found themselves invited into government service ranging from advisory committees (Norman Augustine) even to acting secretary of defense (Patrick Shanahan). The most frequent pattern, however, has been the career service officer who is hired by a defense contractor upon retirement—in genuine appreciation of his or her talents and experience, or to exploit his or her contacts still in government service, or to reward him or her for past favors. Those who have abused the system have, unfortunately, cast a shadow of suspicion over all who partake.

And, of course, the $435 hammers and $640 toilet seats followed the military services into the post–Cold War world, popping up most visibly in the $1,280 self-warming coffee cup that the air force purchased for its flight crews. Senator Charles Grassley challenged the secretary of the air force to explain, but the problem was as old and intractable as the MIC itself. Paul Light notes that this absurdity, like many others, "was an artifact of arcane accounting rules." But, as one DoD official tasked with rooting out such contractor abuses told Congress as far back as 1986, "I think these people cheated, and their attitude is 'we stole it fair and square . . . and we are not going to give it back.'" [16]

The stain of outright fraud continues to sully the DIB. Any commercial enterprise of this size and wealth is bound to breed some unsavory behavior. Anecdotal evidence of particularly odious chicanery continues to circulate in progressive and reformist circles, but still the MIC appears to generate more than its share of scandal. During the Cold War, the "Project on Military Procurement" documented abuses ranging from overpriced spare parts to noncompetitive contract awards.[17] The abuses bear striking resemblance to those reported in 2011 by the under secretary of defense for acquisition, technology and logistics in his "Report to Congress on Contracting Fraud."[18]

No doubt, the list of MIC triumphs since 1991 is long and distinguished, as it was during the Cold War. It includes the F-16 Fighting Falcon fighter/multirole aircraft; the family of unmanned aeronautical vehicles (reconnaissance and armed); precision-guided munitions; aerial and space reconnaissance vehicles; communications interception; data collection and mining; global positioning and mapping; and the Aegis Combat System of ship defense. Against these pillars of America's arsenal must be counted a long list of failed or disappointing weapons systems introduced or operated since the Cold War. Stephen Rodriguez identified 10 egregiously failed weapons acquisitions in the era of the RMA—ranging from the navy's next-generation cruiser and the "Missile Defense Agency's Kinetic Energy Interceptor to the Marine Corps' Expeditionary Fighting Vehicle and the Army's Future Combat System." On the last, the DoD spent almost $53 billion without deploying a single operational system.[19]

The development of new weapons and equipment is a risky business, bound to suffer setbacks and disappointments, but the annual evaluation by the Government Accountability Office suggests that many of the DoD's problems continue to arise from failure to follow best practice.[20] The Federation of American Scientists reported more of the same in 2019.[21] The notion that all US businesses suffer such problems is undermined by the Federal Contractor Misconduct Database. The five leading contractors on the list, by dollar amount of contract awards, are leading corporations in the DIB; in fact, nine of the top 10 are in that category. A closer examination of the total list reveals that only the petroleum and pharmaceutical industries compete with the DIB for total penalties between 1995 and 2018.[22]

Since the end of the Cold War, Congress has continued and expanded acquisitions that neither the DoD nor the White House wanted or recommended. State and local officials have pressured their congressional delegation to launch or sustain DoD undertakings in their locale. Defense contractors lobby multiple members of Congress on behalf of projects with multiple contractors and subcontractors around the country. In many cases, those members of Congress most interested in the undertaking have logrolled their colleagues for votes. The epitome of this kind of abuse may have been Randall Harold "Duke" Cunningham (R, CA), a bona fide navy war hero turned congressman (1991–2005), who accepted bribes from defense contractors to deliver his votes and those of colleagues for projects that the services and the White House often did not want.[23] Probably the most successful effort of any administration to overcome this congressional cronyism was the work of five Base Realignment and Closure Commissions (BRAC, 1988–2005), which required Congress to accept whole packages of closings without singling out exceptions to favor one member of Congress or another.[24]

Finally, the industry practice of salting subcontractors in

multiple and influential states and congressional districts around the country in order to win support of the congressional delegations in those areas continues unabated. One of the perennial arguments for launching or sustaining a project almost anywhere in the United States is jobs. So effective has this practice been that other federal agencies and their contractors have placed their facilities and projects with this distribution in mind. Prime contractors have been similarly mindful in choosing subcontractors.[25] Perhaps the quid pro quo attributed to this relationship is less straightforward than it appears on the surface, but it is nonetheless real.

But the Military-Industrial Complex of the last 30 years also differs in significant ways from that of the Cold War. Perhaps most importantly, the boundaries separating military-civilian and public-private realms have blurred. Since the end of the Cold War, public esteem for the military has risen steadily, even while public approbation for politicians, "experts," and scientists has eroded.[26] The trend reveals itself in the nation's presidents since the end of the Cold War, from the bona fide war hero George H. W. Bush through three presidents who failed to serve in Vietnam—Bill Clinton, George W. Bush (who served at home in the reserves), and Donald Trump—and one president, Barack Obama, who was too young. The number of senators and representatives with histories of military service declined from about three-quarters in the mid-1970s to less than a fifth in the 116th Congress (2019–2021).[27] This drop reflects the changing demography of the armed services since the move to an all-volunteer armed force. It also means, of course, a less experienced and less sympathetic legislature. On the other side of the coin, many members of Congress, most notably Senator Barry Goldwater, continued to serve simultaneously in the military reserves—he rose to be an air force major general—a practice that the Supreme Court has found to be consonant with the Constitutional prohibition against holding two federal offices at once.[28]

The Democratic presidents since the end of the Cold War, Bill Clinton and Barack Obama, did, however, experience some of what political scientist Peter Feaver calls "shirking" from a military establishment generally disdainful of their views of the military and their national security policies. President Clinton had to fire a general officer who reportedly characterized him in public remarks as "gay-loving," "pot-smoking," "draft-dodging," and "skirt-chasing."[29] Obama faced a similar breakdown in principal-agent relations when he recalled and accepted the resignation of General Stanley McChrystal for disrespecting the president in remarks to a reporter for *Rolling Stone*. George W. Bush also suffered public criticisms from serving and retired officers for his policies in Iraq and Afghanistan, and Donald Trump refused to accept the carefully worded resignation of his secretary of defense, retired General James E. Mattis. He insisted the general was already fired for what Trump obviously saw as a failure of loyalty. Public opinion generally followed party affiliation in evaluating these civil-military disputes.

Another sea change in civil-military relations appears to be rising up from the military quagmires in Iraq (2003–2011) and Afghanistan (2001–). These intractable conflicts—fought by volunteer regulars, reservists, National Guard personnel, and contractors—have tested the American military strategy of favoring quality over quantity. Journalist James Fallows lamented in 2015 that Americans "buy weapons that have less to do with battlefield realities than with [their] unending faith that advanced technology will ensure victory, and with the economic interests and political influence of contractors."[30] American fighting men and women have endured repeated tours of duty in Iraq and Afghanistan and come home with life-changing injuries, psychological wounds from anti-insurgent combat, and disenchantment with US foreign policy and the system of government that produced it. Veterans of these wars are running for office at rates not seen for decades, and voters across the US political divide are supporting them enthusiastically. As was the case with post–Cold War presidents and their civil-military relations, the voting public seems better disposed to the veterans than to the elected officials who sent them off to unpopular wars.

Like the boundary between military and civilian realms, the public-private divide between government and civil society has been blurring since the Cold War. The continued turning of the revolving door passes individuals back and forth between the two realms, though perhaps more so in the military than in the intelligence sector. The almost exclusive use of contracts and direct purchase to buy goods and services from the private sector has been loosened by increasing experimentation with joint ventures that regularly violate the old principal-agent relationships. Linda Weiss's *America Inc.?* explores a cornucopia of innovative new relationships negotiated between the government and the private sector. Overcoming obstacles of security classification, conflict of interest, due diligence, and shared liabilities, these new arrangements—or partnerships, as the intelligence community prefers to characterize them—simply identify naturally occurring situations in which the public and private sectors identify win-win undertakings of mutual interest and minimal shared risk. Gone are the days when a director of DARPA is sacked for venturing into the dark and bloody ground of national industrial policy simply because an undertaking smacked of creeping socialism in the minds of some observers. The MIC was always a kind of national industrial policy that could be forgiven if the industry involved was lodged in the DIB.

In the twenty-first century, however, technologies of most interest to the military were being invented and developed by companies in the microelectronics domain, pushed into the marketplace by commercial forces instead of being pulled into existence by military demands. Principal-agent relationships remained, usually in the traditional form of contracts, but new kinds of public-private partnerships were emerging as well. These required new forms of collaborative and cooperative association, and the government was often buying commercial off-the-shelf (COTS) instead of made-to-order. The result was more technology-push, more COTS, more spin-around (and less spin-off), and many more joint ventures. Fred Block and Matthew R. Keller reinforce Weiss's view, arguing that the United States has become a "Developmental Network State," built around hybrid collaborations, a diminished role for large corporations, and increased public funding.[31]

Another major transformation of the Military-Industrial Complex has been the rapid growth and diversification of the intelligence community since 9/11. Now accounting for about 11% of "military" spending, intelligence has focused on real national or homeland defense, as opposed to the force projection of the DoD. The intelligence community has developed its own Intelligence-Industrial Complex (IIC), arguably the greatest US security success of the twenty-first century. In form and function, the IIC resembles the MIC, but it is more self-consciously focused on partnership between the public and private sectors, perhaps because of its concentration on the microelectronics domain. As with the MIC, the principal-client relationship is still palpable, but the working-level institutions are perhaps less formal and more diverse.

Another major change in the new MIC is that military spending—the "burden"—is now far less significant in American national life than it was during the Cold War. To understand the importance of this point, it is necessary to view military spending from several perspectives. If it is viewed in then-year dollars, defense spending rose from $273.3 billion in 1991 to $631.2 billion in 2018, a change of 130%.[32] On a graph (chapter 8, chart 1) this can look like a steeply rising mountain of money, as indeed it is. It explains the alarms raised by such otherwise prudent observers as Fareed Zakaria.[33] But when the same phenomenon is viewed in constant (2017) dollars, the spending is seen to have increased only from $504 billion to $606 billion, a change of just 17%.[34] Instead of spending 5% to 10% of GDP on the military, as the United States did over the course of the Cold War, the country has been spending 3% to 5% since 1991 (chapter 8, chart 2). Remarkably, this dramatic reduction in uninflated US military spending has not significantly undermined America's position in the world. US military spending was about a third of world military spending in 1991; in 2018, it was still about a third. In 2018, the United States spent more on the military than the next seven countries combined, including China, and four or five of those countries were traditional US allies. The United States still has no peer rivals, especially when its allies are weighed in the balance.

But the most dramatic change in the economic "burden" that military spending has imposed on the United States in the post–Cold War world is the percentage of federal spending consumed by the military. How much, in short, does the military take out of the federal budget? To appreciate this change, it is illuminating to go back to 1970, when the US government spent 41.8% of its discretionary budget on "national security" and 38.5% on "human resources." Guns and butter were in something close to equilibrium, even while the US was trying to extricate itself from Vietnam. In 2018, however, almost half a century later, the US spent 15.4% of its discretionary budget on "national security" and 70.5% on "human resources" (chart 5).[35] Those relative priorities hardly seem consistent with alarms about a "National Security State." The Budget Control Act of 2011 notwithstanding, the United States has surely resolved the guns-and-butter argument in favor of butter.

Of course, these statistics on "military spending" fail to capture many aspects of the country's security spending, such as Veteran's Affairs, Homeland Security, Overseas Contingency Operations, interest on the portion of the national debt caused by defense spending, and others. The real cost of the nation's security in 2019 was certainly closer to $1 trillion than it was to the $631 billion of the "base budget.[36] But even taking these more realistic numbers into account, the security "burden" on the country's economy and the federal budget has nonetheless shrunk dramatically since the Cold War.

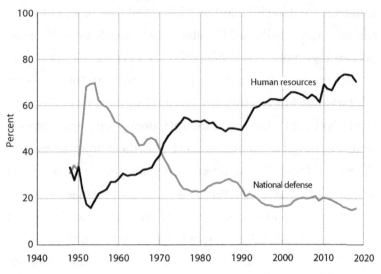

CHART 5. US defense and human resources spending as percentage of federal budget, 1948–2018. *United States, White House, Office of Management and Budget (OMB), Historical Tables, 2019, table 3.1.*

Despite the dominance of US military spending in the world, the once commanding lead of the United States in technological development has also shrunk. Other countries are catching up, sometimes by stealing US secrets more successfully than before. The United States is giving away technology through joint ventures. Competitors can develop cheaper, "good-enough" versions of America's best. Talk of technological parity between the United States and the Soviet Union in the Cold War was mostly rhetorical posturing designed to increase US R&D funding. But the F-35 makes clear that the United States can no longer pretend to be a full generation ahead of its competition.

One of the most important debates among scholars of the MIC has been its impact on the US economy. In the early decades of the Cold War, advocates of military spending emphasized spin-off from military research and development. Critics focused on opportunity costs and what came to be called "crowding out," the diversion of national capital and labor for productive uses. As Eisenhower said:

> Every gun that is made, every warship launched, every rocket fired signifies, in the final sense, a theft from those who hunger and are not fed, those who are cold and are not clothed. This world in arms is not spending money alone. It is spending the sweat of its laborers, the genius of its scientists, the hopes of its children. The cost of one modern heavy bomber is this: a modern brick school in more than 30 cities. It is two electric power plants, each serving a town of 60,000 population. It is two fine, fully equipped hospitals. It is some 50 miles of concrete highway. We pay for a single fighter plane with a half million bushels of wheat. We pay for a single destroyer with new homes that could have housed more than 8,000 people.[37]

A distinguishing characteristic of these early debates is that they were heavily ideological and political, not economic.

Thirty years after the Cold War, the literature on this subject looks very different. Some authors still focus on spin-off or the commercial application of military research and development. One of the most salient advocates is Vernon W. Ruttan, whose *Is War Necessary?* (2006) confronted the issue directly. His answer is an unqualified yes. With a series of case studies ranging from interchangeable parts in the nineteenth century through such Cold War examples as aviation, nuclear energy, computers, the internet, and spaceflight, he highlighted the contributions of defense spending to these "general purpose" or "revolutionary" technologies.[38] Linda Weiss agrees with Ruttan up to a point, but she concentrates on the post–Cold War history, when the term "spin-around" better captured the dynamics of public-private, civil-military cooperation in technologies of the "microelectronics domain." Relying, like Ruttan, largely on anecdotal evidence, Weiss concludes that the "National Security State" has become an "innovation engine" for dual-use technologies.[39]

Skeptics of the proposition that the MIC benefits the national economy have generally taken a very different approach. Finding the theoretical literature on the topic to be fascinating but inconclusive, they focus on empirical studies based on economic analysis of statistical data, more often comparative and international rather than US-centric. Through 2010, these studies in the aggregate concluded that military spending had little impact—perhaps slightly negative—on national economic growth, perhaps helping developed countries somewhat more than developing ones.[40] Some noted that the overall impact of defense spending may differ from the impact of military research and development, on which there is no consensus.[41] By 2013, however, this literature seemed to have coalesced around a stronger consensus that the impact of military spending on national economies was more clearly negative. Indeed, one survey of the literature concluded that the very premise of these studies may have gotten the problem backwards. A more useful conclusion from the research is that "if one wants to have any hope of becoming (militarily) strong, one should invest in one's economy. The best way to security may be through economic growth."[42] Eisenhower could not have said it better. This was his "great equation" in a nutshell, universalized for all the world.

Two distinguished scholars of the MIC who have addressed the same question pursued in this book come to conclusions similar to those presented here, but with a distinctly European perspective. In 2010, J. Paul Dunne and Elisabeth Sköns found an MIC with "a remarkable degree of continuity" with its Cold War predecessor, but an institution "just as pervasive and powerful, more varied, more internationally linked and less visible."[43] Writing in 2019, the same co-authors concluded with shifting emphasis that the MIC was still "just as pervasive and powerful, but considerably less visible, less controllable and more international."[44] In both cases, they paid more attention to the international arms trade and less to US grand strategy and domestic politics. Probably the most important distinction between their interpretations and the one offered here is that the US MIC still appears more powerful on the world stage than it does at home.

In sum, the continuity of the Military-Industrial Complex since the Cold War looks something like the following:

- The United States has a larger security establishment —the DoD and Homeland Security and the Veteran's Administration are the three largest government agencies—but the DIB plays a lesser role.
- An Intelligence-Industrial Complex has emerged within the MIC, accounting for about a tenth of its spending and imitating many—but not all—of its characteristics and practices.
- Military spending is higher in constant dollars and higher still in real dollars, but it is smaller as percentages of US GDP, government spending, and world military spending.

- The American public holds the military in higher regard than during the Cold War. The soldier-statesman Colin Powell has displaced the mad bomber Curtis LeMay as the iconic manager of violence. As a result, the MIC evokes less animus than previously.
- The United States still leads the world in military technology, but the lead has diminished overall and even disappeared in some technologies.
- The United States remains committed to the world's best military technology, but the Pentagon provides a smaller percentage of national R&D funding and relies more on commercial-off-the-shelf (COTS) procurement, spin-around transfer, and cooperative ventures with the private sector, especially in the microelectronics domain.
- The armed services remain committed to legacy weapon platforms (carriers, F-35, etc.), but increasing focus has shifted to C^4ISR technologies from the microelectronics domain.
- The national security sector still relies heavily on contracting, but a higher percentage of that funding goes for services than for goods.
- Contracting for goods and services within the MIC and IIC remains just as beset by waste, fraud, and abuse, but there appears to be even less auditing, management, and enforcement of contracts.

In short, the MIC abides.

ACE. aerospace-computer-electronics

AEC. Atomic Energy Commission

Apollo. NASA moon mission

ARPA. Advanced Research Projects Agency, a.k.a. DARPA

ATF. Advanced Tactical Fighter, concept that became the F-22

B-1. strategic bomber, Valkyrie, a.k.a. "born-again bomber"

bailing out. rescuing a defense contractor facing a financial crisis of its own making

BCA. Budget Control Act

BRAC. Base Realignment and Closure Commission

buying-in. knowingly underbidding a contract expecting to recover short-fall later

C³I. command, control, communication, and information

C⁴ISR. command, control, communication, computers, intelligence, surveillance, reconnaissance

C-5A. scandal-beset Lockheed cargo aircraft that become an icon of LOCUS and cover-up in the 1960s and 1970s

CACI. a professional services and information technology company; formerly Consolidated Analysis Center

capability greed. adding new features to a weapon system during development

CIA. Central Intelligence Agency

COTS. commercial-off-the-shelf

CPSR. Computer Professionals for Social Responsibility

Crusader. army mobile 105 mm howitzer

CSRA. an information technology services company, formed by a 2015

merger of Computer Science Corporation's Government Services
Division and SRA International

DARPA. Defense Advanced Research Projects Agency, a.k.a. ARPA

DDR&E. director of defense research and engineering

demand-pull. technological innovation driven by a perceived need (*see*
technology-push)

DHS. Department of Homeland Security

DIA. Defense Intelligence Agency

DIB. defense industrial base

DNI. director of national intelligence

DoD. Department of Defense

DoE. Department of Energy

dual-use. technologies with military and civilian applications.

Electric Boat. submarine builder, now subsidiary of General Dynamics

ERDA. Energy Research and Development Administration

F-22. Lockheed Martin "Raptor," fifth-generation fighter aircraft

F-35. Lockheed Martin "Lightning II," fifth-generation fighter aircraft

F-111. attack bomber with variable-sweep wings for navy and air force
(*see* TFX)

FFRDC. federally funded research and development center

GAO. Government Accountability Office; Government Accounting Office
before 2004

GDP. gross domestic product

gold-plating. process of adding extravagant features to approved weapon
systems development (*see* capability greed)

Goldwater-Nichols Act. 1986 law that reformed the Department of
Defense

GPS. global positioning system

GVF. government-sponsored VC (venture capital) fund

GWoT. Global War on Terrorism, policy of the George W. Bush administration to seek out terrorists preemptively around the world (*see* War
on Terror)

HDTV. high-definition television

IC. intelligence community

ICBM. intercontinental ballistic missile

IED. improvised explosive device

IIC. Intelligence-Industrial Complex

In-Q-Tel. public-private venture capital firm

integrator. diversified defense contractors with portfolios of complementary companies and capabilities

JSF. Joint Strike Fighter, concept that became the F-35

last supper. 1993 meeting of Deputy Secretary of Defense William Perry with defense industry executives to warn of retrenchment.

LOCUS. late, over cost, and under specifications

LOGCAP. Logistical Civil Augmentation Program

logrolling. colleagues' reciprocal support of pet projects

MAD. mutual assured destruction

mic. generic military-industrial complex

MIC. US Cold War Military-Industrial Complex

microelectronics domain. the field of electronics in which solid-state devices have allowed the development of miniaturized technologies for military and civilian applications

MIMIC. computer research program investigating microwave/millimeter wave monolithic integrated circuits

MIP. Military Intelligence Program (*see* NIP)

MIT. Massachusetts Institute of Technology

MPRI. Military Professional Resources Incorporated

MRAP. Mine-Resistant Ambush Protected vehicle

NASA. National Aeronautics and Space Administration

NATO. North Atlantic Treaty Organization

net-centric warfare. the electronic battlefield; warfare achieving John Boyd's goal of operating within the enemy's Observe-Orient-Decide-Act (OODA) loop; warfare drawing on technologies from the "microelectronics domain"

Newport News Shipbuilding. builder of submarines and aircraft carriers

NGA. National Geospatial-Intelligence Agency

NIP. National Intelligence Program (*see* MIP)

NRC. Nuclear Regulatory Commission

NRE. National Research Establishment

NRO. National Reconnaissance Office

NSA. National Security Agency

NSS. National Security State; or National Security Strategy, a mandated, periodic presidential statement of strategic goals and plans

NTA. National Technology Alliance

OCO. Overseas Contingency Operations

OODA. observe, orient, decide, act

Osprey. a US Marine Corps VTOL/STOL aircraft, V-22

OSRD. Office of Scientific Research and Development

PERT. Program Evaluation and Review Technique

PMF. privatized military firm

PNAC. Project for the New American Century

PSAC. President's Science Advisory Committee

RAND. corporate think tank

revolving door. individuals moving back and forth between senior positions in government and the defense or intelligence industries

RMA. Revolution in Military Affairs

SAIC. Science Applications International Corporation

SALT. Strategic Arms Limitation Talks

SAP. special access programs

SCI. sensitive compartmented information

SDI. Strategic Defense Initiative, a.k.a. Star Wars, President Reagan's plan for ballistic missile defense

Seawolf. nuclear attack submarine class

Sematech. a not-for-profit consortium that derived its name from semiconductor manufacturing technology

spin-around. recent variation of spin-off

spin-off. technology transfer from one realm—usually military—to another—usually private or commercial

SRI. Stanford Research Institute; since 1977 SRI International

STOL. short takeoff and landing

TFX. developmental version of F-111 attack bomber

TSA. Transportation Security Administration

V-22. Osprey VTOL/STOL aircraft

Valkyrie. B-70 bomber

VHSIC. computer research program investigating very high-speed integrated circuits

VTOL. vertical takeoff and landing

Vulcans. informal conservative interest group advocating increased defense spending and more assertive foreign policy

War on Terror. policy of the George W. Bush administration to seek out terrorists preemptively around the world (*see* GWoT)

Introduction

1. In this study, the Military-Industrial Complex (MIC) will denote the American experience since World War II, while military-industrial complex will refer to other comparable institutions in other historical times and places.

2. Alex Roland, *The Military-Industrial Complex*, Historical Perspectives on Technology, Society, and Culture, ed. Pamela Long and Robert C. Post (Washington, DC: American Historical Association, 2001).

3. J. Paul Dunne and Elizabeth Sköns, "The Changing Military-Industrial Complex," Working Paper 1104, Department of Accounting, Economics, and Finance, Bristol Business School, University of the West of England (2011), 1. They suggest that smaller, European versions of the American MIC might better be called "defence industrial networks." See also the same authors' "The Military Industrial Complex," *The Global Arms Trade: A Handbook*, ed. Andrew T. H. Tan (New York: Routledge, 2010), 281–92.

4. Ron Smith, *Military Economics: The Interaction of Power and Money* (New York: Palgrave Macmillan, 2009), 143, 138.

5. One goal of this study is to compare the hostile, canonical literature on the MIC with a smaller, exculpatory body of scholarship that tends to be less anecdotal and more rigorous.

6. The plaintive title of John Stanley Baumgartner's unique defense of the MIC bespeaks the paucity of literature in this vein. See *The Lonely Warriors; Case for the Military-Industrial Complex* (Los Angeles: Nash Pub., [1970]). Some scholars, mostly in article-length studies, have argued that the anecdotal critiques do not stand up to closer scrutiny.

Several such analyses appear in Steven Rosen, ed., *Testing the Theory of the Military-Industrial Complex* (Lexington, MA: Lexington Books, 1973).

7. Article VI of the US Constitution requires public servants to "support" the Constitution itself. The familiar lines from the current oath of office were crafted during the Cold War. Pub. L. 89-554, Sept. 6, 1966, 80 Stat. 424.

8. In 1957, for example, the air force got 47% of the Department of Defense funds allotted to the services, the navy got 27.6%, and the army got 25.3%. US Department of Defense, Office of the Under Secretary of Defense (Comptroller), *National Defense Budget Estimates for FY 2020* (the "Green Book"), May 2019, p.94.

9. See, for example, Joel Davidson, *The Unsinkable Fleet: The Politics of U.S. Navy Expansion in World War II* (Annapolis, MD: Naval Institute Press, 1996).

10. Jacques Ellul, *The Technological Society* (New York: Knopf, 1964).

11. Alex Roland, "Was the Nuclear Arms Race Deterministic?," *Technology and Culture* 51 (Apr. 2010): 444–61.

12. Thomas C. Schelling, "An Astonishing 60 Years: The Legacy of Hiroshima," Nobel Prize Lecture, Stockholm, Sweden, Dec. 8, 2005, p. 1, www.nobelprize.org/uploads2018/06/schelling-lecture,pdf.

13. Smith, *Military Economics*, 100–101.

14. The term "National Security State" gained salience with Daniel Yergen, *Shattered Peace: The Origins of the Cold War and the National Security State* (New York: Houghton Mifflin, 1977); and Michael J. Hogan, *A Cross of Iron: Harry S. Truman and the National Security State, 1945–1954* (New York: Cambridge University Press, 1998).

15. See the differing positions of Michael S. Sherry, *In the Shadow of War: The United States since the 1930s* (New Haven, CT: Yale University Press, 1995); and Aaron L. Friedberg, *In the Shadow of the Garrison State: America's Anti-statism and Its Cold War Grand Strategy* (Princeton, NJ: Princeton University Press, 2000).

16. See, for example, James M. Lindsay, "Parochialism, Policy, and Constituency Constraints: Congressional Voting on Strategic Weapons

Systems," *American Journal of Political Science* 34, no. 4 (Nov. 1990): 936–60.

Chapter 1. Defining the Complex

1. The $1 billion figure comes from Elsbeth E. Freudenthal, *The Aviation Business: From Kitty Hawk to Wall Street* (New York: Vanguard Press, 1940), 35–61. I. B. Holley reports the delivery of aircraft in *Ideas and Weapons: Exploitation of the Aerial Weapon by the United States during World War I; A Study in the Relationship of Technological Advance, Military Doctrine, and the Development of Weapons* (New Haven, CT: Yale University Press, 1953), 106, 131–32.

2. Andrew Gibson and Arthur Donovan, *The Abandoned Ocean: A History of United States Maritime Policy* (Columbia: University of South Carolina Press, 2000), 112–15.

3. Stuart D. Brandes, *Warhogs: A History of War Profits in America* (Lexington: University Press of Kentucky, 1997).

4. Stephen J. Zempolich, "Dwight David Eisenhower and the Military-Industrial Complex: Advocacy to Opposition, 1928–1961" (Senior Honors Thesis, Duke University), 1985.

5. Eisenhower appears to have had in mind what Herbert Hoover called an "associative state." See Ellis Hawley, "Herbert Hoover, the Commerce Secretariat and the Vision of an 'Associative State,' 1921–1928," *Journal of American History* 41 (June 1947): 116–40.

6. John U. Nef, *War and Human Progress: An Essay on the Rise of Industrial Civilization* (Cambridge, MA: Harvard University Press, 1950); Ron Smith, *Military Economics: The Interaction of Power and Money* (New York: Palgrave Macmillan, 2009), 137.

7. Alex Roland, *Underwater Warfare in the Age of Sail* (Bloomington: Indiana University Press, 1978).

8. Elting Morison, *Men, Machines, and Modern Times* (Cambridge, MA: MIT Press, 1966), 17–44.

9. John Ellis, *The Social History of the Machine Gun* (New York: Pantheon Books, 1975).

10. David E. Johnson, *Fast Tanks and Heavy Bombers: Innovation in the U.S. Army, 1917–1945* (Ithaca, NY: Cornell University Press, 1998).

11. Some of these motives are explored in Michael Sherry, *Preparing for the Next War: American Plans for Postwar Defense, 1941–1945* (New Haven, CT: Yale University Press, 1977). Vannevar Bush's plan for a National Research Establishment, advocated in *Science, the Endless Frontier: A Report to the President* (Washington, DC: Government Printing Office, 1945), is discussed in this volume's conclusion.

12. Interview of General Andrew J. Goodpaster, Jr., by Eugene M. Emme and Alex Roland, the Pentagon, July 22, 1974, NASA Historical Reference Collection, File Number 13469; Richard V. Damms, "Containing the Military-Industrial-Congressional Complex: President Eisenhower's Science Advisers and the Case of the Nuclear-Powered Aircraft," *Essays in Economic and Business History* 14 (1996): 279–89; James Ledbetter, *Unwarranted Influence: Dwight Eisenhower and the Military-Industrial Complex* (New Haven, CT: Yale University Press, 2011); Paul C. Light, *The Government-Industrial Complex: The True Size of the Federal Government, 1984–2018* (New York: Oxford University Press, 2019), 180, n. 16.

13. Stuart W. Leslie, *The Cold War and American Science: The Military-Industrial-Academic Complex at MIT and Stanford* (New York: Columbia University Press, 1993); Rebecca S. Lowen, *Creating the Cold War University: The Transformation of Stanford* (Berkeley: University of California Press, 1997).

14. C. Wright Mills, *The Power Elite* (New York: Oxford University Press, 1959).

15. "Military-Industrial Complex Speech, Dwight D. Eisenhower, 1961," *The Avalon Project: Documents in Law, History and Diplomacy* (2008), https://avalon.law.yale.edu/20th_century/eisenhower001.asp.

16. Delores E. Janiewski, "Eisenhower's Paradoxical Relationship with the 'Military-Industrial Complex,'" *Presidential Studies Quarterly*, 4 no. 41 (Dec. 2011): 667–93; Gene M. Lyons and Louis Morton, "School for Strategy," *Bulletin of the Atomic Scientists* 17, no. 3 (1961), 103–6, at 105.

17. Gordon Adams, *The Politics of Defense Contracting: The Iron Triangle* (New Brunswick, NJ: Transaction Books, [1981] 1982).

18. James McCartney, with Molly Sinclair McCartney, *America's War Machine: Vested Interests, Endless Conflicts* (New York: St. Martin's, 2015), 44–45, quote at 45. The Eisenhower Library has reportedly found no evidence supporting this story in the twenty or so drafts of the speech in its archive.

19. As this literature appeared in the 1970s, Robert D. Cuff observed that it was "an intensely ideological body of work." Robert D. Cuff, "An Organizational Perspective on the Military-Industrial Complex," *Business History Review* 52 (Summer 1978): 251.

20. Mark Twain and Charles Dudley Warner, *The Gilded Age: A Novel* (London: George Routledge, 1874).

21. He is often misquoted (even in the 2001 edition of this book) as saying, "What is good for General Motors is good for America." At his confirmation hearings in January 1953, he actually said, "for years I thought that what was good for our country was good for General Motors, and vice versa." Ellen Terrell, "When a Quote Is Not (Exactly) a Quote: General Motors," *Inside Adams: Science, Technology, and Business*, Apr. 22, 2016, blogs.loc.gov/inside_adams/2016/04/when-a-quote-is-not -exactly-a-quote-general-motors/.

22. Richard Cowen, "Iron and an Early Military-Industrial Complex," *History of Life* (New York: McGraw Hill, 1976).

23. Paul A. C. Koistinen, *The Military-Industrial Complex: A Historical Perspective* (New York: Praeger, 1980). See also the first three volumes of the same author's pentalogy: *Beating Plowshares into Swords: The Political Economy of American Warfare, 1606–1865*; *Mobilizing for Modern War: The Political Economy of American Warfare, 1865–1919*; and *Planning War, Pursuing Peace: The Political Economy of American Warfare, 1920–1939*, all published by the University Press of Kansas— 1996, 1997, and 1998, respectively. See also Frank B. Cooling, *Gray Steel and Blue Water Navy: The Formative Years of America's Military-Industrial Complex* (Hamden, CT: Archon Books, 1979).

24. Katherine Epstein, *Torpedo: Inventing the Military-Industrial*

Complex in the United States and Great Britain (Cambridge, MA: Harvard University Press, 2014); Kurt Hakemer, *The U.S. Navy and the Origins of the Military-Industrial* Complex (Annapolis, MD: Naval Institute Press, 2002); Philip Macdougall, *Chatham Dockyard: The Rise of a Military-Industrial Complex* (Cheltenham, UK: History Press, 2012); Masako Ikegami-Anderson, *The Military-Industrial Complex: The Cases of Sweden and Japan* (Aldershot, UK: Dartmouth Publishing, 1992); Mark Harrison, *Soviet Industry and the Red Army under Stalin: A Military-Industrial Complex?*, Economic Research Papers, no. 610 (Coventry, UK: University of Warwick, 2001). David Rohde suggests that the United States has a "security-industrial complex" built around a "culture of secrecy." "The Security-Industrial Complex," *The Atlantic*, June 15, 2013. J. Paul Dunne and Elisabeth Sköns speak of "defense industrial networks" in "The Changing Military-Industrial Complex," paper presented at School of Economics, University of West England, Economics Series, March 2011, http://carecon.org.uk/DPs/1104.pdf, p. 6.

25. William H. McNeill, *The Pursuit of Power: Technology, Armed Force, and Society since 1000 A.D.* (Chicago: University of Chicago Press, 1982), 269–306.

26. C. Vann Woodward, "The Age of Reinterpretation," *American Historical Review* 66 (1960): 1–19.

27. Daniel Yergen, *Shattered Peace: The Origins of the Cold War and the National Security State* (New York: Houghton Mifflin, 1977); Michael J. Hogan, *A Cross of Iron: Harry S. Truman and the National Security State, 1945–1954* (Cambridge, MA: Cambridge University Press, 1998).

28. Michael H. Armacost, *The Politics of Weapons Innovation: The Thor-Jupiter Controversy* (New York: Columbia University Press, 1969).

29. George A. Reed, "U.S. Defense Policy, U.S. Air Force Doctrine, and Strategic Nuclear Weapon Systems, 1958–1964: The Case of the Minuteman ICBM" (PhD dissertation, Duke University, 1986).

30. ARPA was renamed DARPA, the Defense Advanced Research Projects Agency, in 1973. It reverted to ARPA in 1993 and back to DARPA in 1996.

"ARPA Becomes DARPA," accessed Jan. 24, 2020, www.darpa.mil/about
-us/timeline/arpa-name-change. See also Sharon Weinberger, *The
Imagineers of War: The Untold Story of DARPA, the Pentagon Agency that
Changed the World* (New York: Knopf, 2017), 203–204.

31. Walter A. McDougall, . . . *the Heavens and the Earth: A Political
History of the Space Age* (New York: Basic Books, 1985).

32. *Project Horizon: A U.S. Army Study for the Establishment of a
Lunar Military Outpost,* 2 vols. (Washington, DC: United States Army,
June 8, 1959).

33. *Aeronautics and Space Report of the President* (Washington, DC:
NASA, 2017), 215–16.

34. *Aeronautics and Space Report of the President* (Washington, DC:
NASA, 2004), 10. For an overview of what was happening clandestinely,
see William E. Burrows, *Deep Black: Space Espionage and National
Security* (New York: Random House, 1986).

35. Alain C. Enthoven and K. Wayne Smith, *How Much Is Enough?
Shaping the Defense Program, 1961–1969* (New York: Harper & Row,
1971).

36. Valerie L. Adams, *Eisenhower's Fine Group of Fellows: Crafting a
National Security Policy to Uphold the Great Equation* (Lanham, MD:
Lexington Books, 2006).

37. Robert R. Bowie and Richard H. Immerman, *Waging Peace: How
Eisenhower Shaped an Enduring Cold War Strategy* (New York: Oxford
University Press, 1998), 75, 96–108.

38. "The Fortune Directory: The 500 Largest U.S. Industrial Corpora-
tions," *Fortune* (July 1959): 12–16; William Proxmire, "Retired High-
Ranking Military Officers," in Carroll W. Pursell, ed., *Military Industrial
Complex* (New York: Harper & Row, 1972), 260–62.

39. H. L. Nieburg, *In the Name of Science* (Chicago: Quadrangle Books,
1966), 184–99.

40. Seymour Melman, *Pentagon Capitalism: The Political Economy of
War* (New York: McGraw-Hill, 1970), 227.

41. Linda R. Cohen, Roger G. Noll, Jeffrey S. Banks, Susan A. Edelman,
and William M. Pegram, *The Technology Pork Barrel* (Washington, DC:

Brookings Institution, 1991). These case studies illuminate civilian analogs of the Military-Industrial Complex.

42. Kenneth R. Mayer, *The Political Economy of Defense Contracting* (New Haven, CT: Yale University Press, 1991), 3–10.

43. Ann Markusen, Peter Hall, Scott Campbell, and Sabina Deitrick, *The Rise of the Gunbelt: The Military Remapping of Industrial America* (New York: Oxford University Press, 1991).

44. James M. Lindsay, "Parochialism, Policy, and Constituency Constraints: Congressional Voting on Strategic Weapons Systems," *American Journal of Political Science* 34, no. 4 (Nov. 1990): 936–60; Mayer, *Political Economy of Defense Contracting*; Barry S. Rundquist and Thomas M. Casey, *Congress and Defense Spending: The Distributive Politics of Military Procurement* (Norman: University of Oklahoma Press, 2002); Rebecca U. Thorpe, *The American Warfare State: The Domestic Politics of Military Spending* (Chicago: University of Chicago Press, 2014). Thorpe stresses the impact of contractors in small communities in the sunbelt, not necessarily the gunbelt.

45. Nick Kotz, *Wild Blue Yonder: Money, Politics, and the B-1 Bomber* (New York: Pantheon Books, 1988).

46. Also called "low-balling." British economist Ron Smith calls it a "conspiracy of optimism." Smith, *Military Economics*, 112.

47. Arnold Levine, *Managing NASA in the Apollo Era*, NASA SP-4102 (Washington, DC: NASA, 1982), 155.

48. Alex Roland, "The Shuttle: Triumph or Turkey?," *Discover* 6 (Nov. 1985): 29–49.

49. Michael Sherry argues that it did in *In the Shadow of War*, countered by Aaron L. Friedman, *In the Shadow of the Garrison State*.

50. Harold Lasswell, "The Garrison State," *The American Journal of Sociology* 46 (Jan. 1941): 455–68; reprinted in Jay Stanley, ed., *Essays on the Garrison State* (New Brunswick, NJ: Transaction Publishers, 1997), 55–75.

51. Lasswell, "Garrison State," 59–60.

52. Alex Roland, "The Grim Paraphernalia: Eisenhower and the

Garrison State," in Dennis Showalter, ed., *Forging the Shield: Eisenhower and National Security in the 21st Century* (Carson City, NV: Imprint Publications, 2005), 13–22.

53. Eisenhower said in a press conference on March 11, 1959, that continuing increases in defense spending threatened to drive the United States into a "garrison state." "The President's News Conference," *The American Presidency Project*, accessed Nov. 2, 2020, https://www.presi dency.ucsb.edu/documents/the-presidents-news-conference-226.

54. James Barron, "High Cost of Military Parts," *New York Times*, Sept. 1, 1983, p. 1, www.nytimes.com/1983/09/01/business/high-cost-of-mili tary-parts.html. For a thoughtful discussion of such overruns, see Jacques Gansler, *Affording Defense* (Cambridge, MA: MIT Press, 1991), 195–207.

55. Frank Gibney, "The Missile Mess," *Harper's Magazine* (Jan. 1, 1960), 38–45; Edmund Beard, *Developing the ICBM: A Study in Bureau-cratic Politics* (New York: Columbia University Press, 1976); Reed, "U.S. Defense Policy, U.S. Air Force Doctrine, and Strategic Nuclear Weapon Systems, 1958–1964."

56. Blair Haworth introduced the concept of "capability greed" in his study of *The Bradley and How It Got That Way: Technology, Institutions, and the Problem of Mechanized Infantry in the United States* (Westport, CT: Greenwood Press, 1999).

57. Quoted in James Fallows, "The Military-Industrial Complex," *Foreign Affairs*, 133. (Nov.–Dec. 2002), 47.

Chapter 2. Civil-Military Relations

1. Samuel P. Huntington, *The Soldier and the State: The Theory and Politics of Civil-Military Relations* (Cambridge, MA: Belknap Press of Harvard University Press, 1957).

2. James Fallows, *National Defense* (New York: Random House, 1981), 10.

3. Robert H. Johnson, *Improbable Dangers: U.S. Conceptions of Threat in the Cold War and After* (New York: St. Martin's, 1997).

4. Robert F. Coulam, *Illusions of Choice: Problems in the Develop-

ment of F-111 Fighter-Bomber, Teaching and Research Materials, no. 14 (Cambridge, MA: Public Policy Program, John F. Kennedy School of Government, 1973), 180–83.

5. Robert J. Art, *The TFX Decision: McNamara and the Military* (Boston: Little, Brown, 1968), 44.

6. Donald MacKenzie, *Inventing Accuracy: An Historical Sociology of Nuclear Missile Guidance* (Cambridge, MA: MIT Press, 1990).

7. Quoted in *The Military Critical Technologies List*, ADA-146 998 (Washington, DC: Office of the Under Secretary of Defense Research and Engineering, Oct. 1984), 1.

8. *Military Critical Technologies*, A-i.

9. Department of Defense, "Critical Technologies Plan," for the Committees on Armed Services, United States Congress, AD-A234 900, May 1991, quote at II-5.

10. Herbert N. Foerstel, *Secret Science: Federal Control of American Science and Technology* (Westport, CT: Praeger, 1993), 22.

11. "Secrecy in University-Based Research: Who Controls? Who Tells?," special issue of *Science, Technology, and Human Values* 10 (Spring 1982); *Physics Today* (1985), passim.

12. Daniel Patrick Moynihan, *Secrecy: The American Experience* (New Haven, CT: Yale University Press, 1998).

13. Foerstel, *Secret Science*, 13.

14. "National Critical Technologies List," Appendix A, Mar. 7, 2000, www.whitehouse.gov/WH/EOP/OSTP/CTIformatted/AppA/appa.html, p. 1.

15. Huntington, *Soldier and the State*, 11–14.

16. Harvey Sapolsky, *The Polaris System Development: Bureaucratic and Programmatic Success in Government* (Cambridge, MA; Harvard University Press, 1972).

17. John Clayton Lonnquest, "The Face of Atlas: General Bernard Schriever and the Development of the Atlas Intercontinental Ballistic Missile, 1953–1960" (PhD dissertation, Duke University, 1996).

18. Stephen B. Johnson, "Samuel Phillips and the Taming of Apollo," *Technology and Culture* 42, no. 4 (Oct. 2001): 685–709.

Chapter 3. State and Industry

1. Merritt Roe Smith, *Harpers Ferry Armory and the New Technology: The Challenge of Change* (Ithaca, NY: Cornell University Press, 1977).

2. David Hounshell, *From the American System to Mass Production, 1800–1932: The Development of Manufacturing Technology in the United States* (Baltimore: Johns Hopkins University Press, 1984).

3. Vannevar Bush, *Science, the Endless Frontier: A Report to the President* (Washington, DC: Government Printing Office, July 1945).

4. G. Pascal Zachary, *Endless Frontier: Vannevar Bush, Engineer of the American Century* (New York: Free Press, 1997).

5. Peter J. Westwick, *The National Labs: Science in an American System, 1947–1974* (Cambridge, MA: Harvard University Press, 2003).

6. Michael Hiltzik, *Big Science: Ernest Lawrence and the Invention that Launched the Military-Industrial Complex* (New York: Simon & Schuster, 2015).

7. This theme is explored in Alex Roland, *Model Research: The National Advisory Committee for Aeronautics, 1915–1958*, 2 vols. (Washington, DC: NASA, 1985).

8. Lonnquest, "The Face of Atlas," 87–92; H. L. Nieburg, *In the Name of Science*, rev. ed. (Chicago: Quadrangle Books, 1970), 200–17.

9. For a contrasting story of success, see Harvey M. Sapolsky, *Polaris System Development: Bureaucratic and Programmatic Success in Government* (Cambridge, MA: Harvard University Press, 1972).

10. Thomas P. Hughes, "The Evolution of Large Technological Systems," in Wiebe E. Bijker, Thomas P. Hughes, and Trevor J. Pinch, eds., *The Social Construction of Technological Systems: New Directions in the Sociology and History of Technology* (Cambridge, MA: MIT Press, 1990), 51–82. Hughes used the Atlas program as an example of "massive research and development projects" in Thomas P. Hughes, *Rescuing Prometheus* (New York: Pantheon Books, 1995), 69–139.

11. Nora Simon and Alain Minc, *The Computerization of Society: A Report to the President of France* (Cambridge, MA: MIT Press, 1980).

12. Ron Smith, *Military Economics: The Interaction of Power and Money* (New York: Palgrave Macmillan, 2009), 167; Jacques S. Gansler, *Democracy's Arsenal: Creating a Twenty-First Century Defense Industry* (Cambridge, MA: MIT Press, 2011), 283–86.

13. But note the critique of this view in Eugene Gholz, "Eisenhower versus the Spin-off Story: Did the Rise of the Military–Industrial Complex Hurt or Help America's Commercial Aircraft Industry?," *Enterprise & Society* 12, no. 1 (Mar. 2011): 46–95.

14. See Ronald O'Rouke, *V-22 Osprey Tilt-Rotor Aircraft: Background and Issues for Congress*, CRS Report RL 31384, June 10, 2019, for the mixed reviews on this controversial aircraft.

15. Berkeley Rice, *The C-5A Scandal: An Inside Story of the Military-Industrial Complex* (Boston: Houghton Mifflin, 1971).

16. Quoted in William D. Hartung, *Prophets of War: Lockheed Martin and the Making of the Military-Industrial Complex* (New York: Nation Books, 2011), 71–72.

Chapter 4. Among Government Agencies

1. Most famously, the Reagan administration secretly sold arms to Saddam Hussein and to Iran during the Iran-Iraq War. The US also sold arms to Turkey and to Kurdish rebels in Turkey, where the two sides opposed each other in 2019. William Hartung, "Turkey's Invasion of Syria, Made in the U.S.A.," *Forbes*, Oct. 10, 2019), https://www.forbes.com/sites/williamhartung/2019/10/10/turkeys-invasion-of-syria-made-in-the-usa/#2d3ca0525483.

2. Jonathan Turley, quoted in David Armstrong, "The Nation's Dirty, Big Secret," *Boston Globe*, Nov. 14, 1999, http://www/boston.com/globe/nation/packages/pollution/day1.htm. Of course, civilian agencies, from the US Mint to the National Parks, have also run afoul of EPA regulations. As one environmental monitor put it, "the government remains the nation's premier environmental felon."

3. Federation of American Scientists, Intelligence Resource Program,

"Intelligence Budget Data," accessed June 27, 2020, https://fas.org/irp/budget/.

4. Walter A. McDougall, . . . *the Heavens and the Earth: A Political History of the Space Age* (New York: Basic Books, 1985).

5. Paul A. David, "Path Dependence: A Foundational Concept for Historical Social Science," *Cliometrica* (July 2007): 91–114.

6. Spencer Weart, *Nuclear Fear: A History of Images* (Cambridge, MA: Harvard University Press, 1988); Weart, *The Rise of Nuclear Fear* (Cambridge, MA: Harvard University Press, 2012).

Chapter 5. The Scientific-Technical Community

1. "Military-Industrial Complex Speech, Dwight D. Eisenhower, 1961," *The Avalon Project: Documents in Law History and Diplomacy,* accessed Dec. 9, 2019, http://www.avalon.law.yale.edu/20th_century/eisenhower 001.asp.

2. Stuart W. Leslie, *The Cold War and American Science: The Military-Industrial-Academic Complex at MIT and Stanford* (New York: Columbia University Press, 1993); 2–41, 90–99.

3. "The Top 200 Defense Contractors," special issue of *Military Forum* 6 (August 1989): 67.

4. Leslie, *Cold War and American Science,* 63–75.

5. See the work of Paul Forman and Paul Edwards and further discussion about the misdirection of research funds in chapter 6.

6. Leslie, *Cold War and American Science,* 252. It should be noted that funding from the National Institutes of Health had a similar impact on university medical centers around the country.

7. Leslie, *Cold War and American Science,* 243–47.

8. Sarah Bridger, *Scientists at War: The Ethics of Cold War Weapons Research* (Cambridge, MA: Harvard University Press, 2015).

9. Jonathan Schell, *The Fate of the Earth* (New York: Knopf, 1982).

10. Jessica Wang, *American Science in an Age of Anxiety: Scientists,*

Anticommunism, and the Cold War (Chapel Hill: University of North Carolina Press, 1999).

11. See chapter 6, pp. 77–79 below.

12. Peter Goodchild, *Edward Teller: The Real Dr. Strangelove* (Cambridge, MA: Harvard University Press, 2004).

13. Quoted in Fred J. Cook, *The Warfare State* (New York: Macmillan, 1962), 243. See also William J. Broad, *Teller's War: The Top-Secret Story behind the Star Wars Deception* (New York: Simon & Schuster, 1992).

14. Stephen Shapin, "Megaton Man," review of Teller's *Memoirs* in *London Review of Books*, Apr. 25, 2002, p. 7, http://www.scholar.har vard.edu/files/shapin/files/shapin_lrbteller.pdf. Shapin reported that he had this information from Herbert York, Teller's former boss, who had it from an interview of Eisenhower after his presidency. York said that Eisenhower also named Wernher von Braun.

Chapter 6. Society and Technology

1. See Ann Douglas's thoughts on "cold-war speak" and other cultural features of the era in Ann Douglas, "Periodizing the American Century: Modernism, Postmodernism, and Postcolonialism in the Cold War Context," *Modernism/Modernity* (1998): 71–98, quote at 79.

2. Albert Wohlstetter, "The Delicate Balance of Terror," *Foreign Affairs* 37 (Jan. 1959): 211–34.

3. Mary Kaldor, *The Baroque Arsenal* (New York: Hill & Wang, 1981), 186–87.

4. Eugene Burdick and Harvey Wheeler, *Fail-Safe* (New York: McGraw-Hill, 1962).

5. Jacques Ellul, *The Technological Society* (New York: Knopf, 1964).

6. Lewis Mumford, *The Myth of the Machine*, 2 vols. (New York: Harcourt Brace Jovanovich, 1967–1970).

7. Langdon Winner, *Autonomous Technology: Technics out of Control as a Theme in Political Thought* (Cambridge, MA: MIT Press, 1977).

8. For a critique of technological determinism, see Alex Roland, "Is Military Technology Deterministic?," *Vulcan* 7 (2019): 19–33.

9. Paul Forman, "Behind Quantum Electronics: National Security as Basis for Physical Research in the United States, 1940–1960," *Historical Studies in the Physical Sciences*, 18 (1987): 149–229; Paul Forman and José M. Sánchez-Ron, eds., *National Military Establishments and the Advancement of Science and* Technology (Dordrecht, Netherlands: Kluwer Academic Publishers, 1996), 9–14, 261–326.

10. Serious attempts to count the cost of the US nuclear arsenal appear in Stephen I. Scwhartz, ed., *Atomic Audit: The Costs and Consequences of U.S. Nuclear Weapons since 1940* (Washington, DC: Brookings Institution Press, 1998); Stephen I. Schwartz with Deepti Choosey, *Nuclear Security Spending: Assessing Costs, Examining Priorities* (Washington, DC: Carnegie Endowment for International Peace, 2009).

11. Paul Higgs argues that the Military-Industrial Complex clearly shaped the American economy in the second half of the twentieth century, but he adds that economic historians need to learn more about its impact on "the rate and direction of technological change." Paul Higgs, "The Cold War Economy: Opportunity Costs, Ideology, and the Politics of Crisis," *Explorations in Economic History* 31 (1994): 283–312.

12. Paul Edwards, *The Closed World: Computers and the Politics of Discourse in Cold War America* (Cambridge, MA: MIT Press, 1996).

13. Since the Cold War, C³I has become C⁴ISR, command, control, communications, computers, intelligence, surveillance, and reconnaissance. See chapter 9.

Chapter 7. International Arms Trade

1. A good Cold War introduction is Anthony Sampson, *The Arms Bazaar: From Lebanon to Lockheed* (New York: Viking, 1977). Mary Kaldor, *The Baroque Arsenal* (New York: Wang/Farrar, Straus & Giroux, 1981) concentrates more on the weapons than the merchants. Reliable longitudinal data appear in Stockholm International Peace Research

Institute (SIPRI) "Arms Transfer Database" (1950), http://www.sipri
.org/databases/armstransfers.

2. Aaron Karp, "The Rise of Black and Gray Markets," *The ANNALS of
the American Academy of Political and Social Science* 535 (1994): 175–89.

3. Federation of American Scientists, "Status of World Nuclear Forces,"
accessed Jan. 22, 2010, http://www.fas.org/programs/ssp/nukes/nuclear
weapons/nukestatus.html. North Korea signed the nonproliferation
treaty but withdrew after the Cold War to pursue its own nuclear weapons
program.

4. Catherine Collins and Douglas Frantz, "The Long Shadow of A. Q.
Khan: How One Scientist Helped the World Go Nuclear," *Foreign Affairs*,
Jan. 31, 2018, https://www.foreignaffairs.com/articles/north-korea
/2018-01-31/long-shadow-aq-khan.

5. Andrew J. Pierre, ed., *Cascade of Arms: Managing Conventional
Weapons Proliferation* (Washington, DC: Brookings Institution Press,
1997).

6. Alex Roland, "Keep the Bomb," *Technology Review* (Aug.–Sept.
1995): 67–69.

7. *Report from Iron Mountain on the Possibility and Desirability
of Peace* (New York: Dial Press, 1967), a hoax that purported to be a
purloined copy of a secret US government document proving that the
economy would collapse without the arms industry, stirred great interest
in the Vietnam era, but "The Effect of Foreign Military Sales on the U.S.
Economy" (Washington, DC: Congressional Budget Office, 1976) found
that elimination of foreign military sales would reduce US GDP less
than 1%.

8. Seymour Melman, *Pentagon Capitalism: The Political Economy of
War* (New York: McGraw-Hill, 1970). Compare Congressional Budget
Office Staff Working Paper, *The Effect of Foreign Military Sales on the
U.S. Economy* (Washington, DC: Congress of the United States, July 23,
1976).

9. At a fundamental level, many of these transactions entailed a
transfer of wealth from the taxpayers of the selling states to their own
arms manufacturers.

10. Jurgen Brauer and J. Paul Dunne, eds., *Arms Trade and Economic Development: Theory, Policy, and Cases in Arms Trade Offsets* (London: Routledge, 2004), 1–2.

11. Edward J. Laurance, *The International Arms Trade* (New York: Lexington Books, 1992), 43, 152; William F. Sater and Holger H. Herwig, "The Art of the Deal," in Donald J. Stoker and Jonathan A. Grant, eds., *Girding for Battle: The Arms Trade in a Global Perspective, 1815–1940* (Westport, CT: Praeger, 2003), 53–96.

12. Graham Allison, *Nuclear Terrorism: The Ultimate Preventable Catastrophe* (New York: Times Books / Henry Holt, 2004).

13. Michael Klare, "The Subterranean Arms Trade: Black-Market Sales, Covert Operations and Ethnic Warfare," in Andrew J. Pierre, ed., *Cascade of Arms: Managing Conventional Weapons Proliferation* (Washington, DC: Brookings Institution Press, 1997), 43–71.

14. Michael Mihalka, "Supplier-Client Patterns in Arms Transfers: The Developing Countries, 1967–1976," in Stephanie G. Neuman and Robert E. Harkavy, eds., *Arms Transfers in the Modern World* (New York: Praeger, 1979), 60.

15. Attempts were made by three organizations to track world military expenditures, including armaments, through much of the Cold War and beyond: the Stockholm International Peace Research Institute's SIPRI *Yearbook* (annual 1969–present), the United States Arms Control and Disarmament Agency's *World Military Expenditures and Arms Transfers* (title varies, 1961–1993), and, less comprehensively, Britain's International Institute for Strategic Studies' *The Military Balance* (1959–present). See Edward T. Fei, "Understanding Arms Transfers and Military Expenditures: Data Problems," in Stephanie G. Neuman and Robert E. Harkavy, eds., *Arms Transfers in the Modern World* (New York: Praeger, 1979), 37–46.

16. Keith Krause, *Arms and the State: Patterns of Military Production and Trade* (Cambridge: Cambridge University Press, 1992), 86, 87, 133.

17. "Trends in global export volume of trade in goods from 1950 to 2018," accessed Jan. 6, 2020, www.statista.com/statistics/264682/world wide-export-volume-in-the-trade-since 1950/.

Chapter 8. New World Order

1. Bush and his National Security Advisor Brent Scowcroft argued for the former in George H. W. Bush and Brent Scowcroft, *A World Transformed* (New York: Knopf, 1998). Woodrow Wilson used similar language after World War I to promote his Fourteen Points, including the League of Nations. For example, he invoked a "new order of the world" in "Address at the University of Minnesota Armory in Minneapolis," The American Presidency Project, Sept. 9, 1919, www.presidency.ucsb.edu /documents/address-the-university-minnesota-armory-minneapolis.

2. Alex Roland, "The Grim Paraphernalia: Eisenhower and the Garrison State," in Dennis Showalter, ed., *Forging the Shield: Eisenhower and National Security in the 21st Century* (Carson City, NV: Imprint Publications, 2005) 13–22.

3. Federation of American Scientists, "Russian Military Budget," accessed Dec. 11, 2019, fas.org/nuclear/guide/Russia/agency/mo-budget .htm. Soviet/Russian budget figures for the military were notoriously unreliable, even before the collapse of the Soviet Union. These FAS figures offer a rough approximation of what was happening during this period in the country's history based on US State Department estimates in 2000.

4. "Defense Planning Guidance, FY 1994–1999," Apr. 16, 1992, https:// www.archives.gov/files/declassification/iscap/pdf/2008-003-docs1-12 .pdf. This document has an original draft, comments, and a draft of the final version dated April 16, 1992. That draft employs the circumlocution that the United States will block "the reemergence of a global threat to the interests of the United States." That is, no peer rival.

5. "Excerpts from 1992 Draft "Defense Planning Guidance," *The War Behind Closed Doors*, July 22, 2019, www.pbs.org/wgbh/pages/frontline /shows/iraq/etc/wolf.html.

6. Patrick E. Tyler, "U.S. Strategy Plan Calls for Insuring No Rivals Develop," *New York Times* (1923-Current file); Mar. 8, 1992; Pro Quest Historical Newspapers, *New York Times*, 1.

7. John Lewis Gaddis, "A Grand Strategy of Transformation," *Foreign Policy* (Nov./Dec. 2002): 50–57; Walter A. McDougall claims that

Presidents Clinton, George W. Bush, and Barack Obama all embraced
"preponderance." *The Tragedy of American Foreign Policy: How America's
Civil Religion Betrayed the National Interest* (New Haven, CT: Yale
University Press, 2016), 18.

8. Terrence Ball and Richard Daggar, "Neoconservatism," *Encyclopedia Britannica*, accessed Aug. 7, 2019, www.britannica.com/topic/neo
conservatism.

9. Economy Watch, accessed on July 9, 2019, economywatch.com
/economic-statistics/year/1992; "Gross World Product," en.wikipedia.org
/wiki/Gross_world_product; "US GDP by Year Compared to Recessions
and Events," accessed July 9, 2019, https://thebalance.com/us-gdp-by
-year-3305543.

10. United States, White House, Office of Management and Budget
[OMB], *Historical Tables*, table 3.1, Outlays by Superfunction and
Function: 1940–2024, accessed Oct. 21, 2019, www.whitehouse.gov/omb
/historical-tables/. In constant dollars, the decline in defense spending
actually began gently in 1985. Brad Plumer, "America's Staggering
Defense Budget, in Charts," *Washington Post*, Jan. 7, 2013, www.wash
ingtonpost.com/news/wonk/wp/2013/01/07/everything-chuck-hagel
-needs-to-know-about-the-defense-budget-in-charts/.

11. Alex Mintz, ed., "Guns versus Butter: A Disaggregated Analysis" in
The Political Economy of Military Spending in the United States (London:
Rutledge, 1992), 185–95.

12. Ron Smith, *Military Economics: The Interaction of Power and
Money* (New York: Palgrave Macmillan, 2009), 159.

13. Helen Caldicott, *The New Nuclear Danger: George W. Bush's
Military-Industrial Complex* (New York: New Press, 2002).

14. Stephen I. Schwartz with Deepti Choosey, *Nuclear Security
Spending: Assessing Costs, Examining Priorities* (Washington, DC:
Carnegie Endowment for International Peace, 2009), 7–8.

15. See chapter 1.

16. Sam Seitz, "The Origins and Effects of the Second Offset," *Politics
in Theory and Practice*, Oct. 1, 2019, www.politicstheorypractice.com
/2019/10/01/the-origins-and-effects-of-the-second-offset/; James Hasik,

"Beyond the Third Offset: Matching Plans for Innovation to a Theory of Victory," *Joint Forces Quarterly* 91, no. 4 (4th quarter 2018): 14–21.

17. Daniel Ford, *A Vision So Noble: John Boyd, the OODA Loop, and America's War on Terror* (Durham, NH: Warbird Books, 2010).

18. Paul Dickson, *The Electronic Battlefield* (Bloomington: Indiana University Press, 1976).

19. Gary Hart and William S. Lind, *America Can Win: The Case for Military Reform* (Bethesda, MD: Adler & Adler, 1986). Its limited effectiveness is addressed in Jeffrey Record, "The Military Reform Caucus," *Washington Quarterly* 6, no. 2 (1983): 125–29.

20. J. David Bolter, *Turing's Man: Western Culture in the Computer Age* (Chapel Hill: University of North Carolina Press, 1984).

21. Walter Millis similarly claimed that "the one great, determining factor which shaped the course of the Second World War" was the internal combustion engine. Walter Millis, *Arms and Men: A Study in American Military History* (New Brunswick, NJ: Rutgers University Press, [1956] 1981), 283.

22. The same search for a defining military technology of the twenty-first century informed my earlier essay: Alex Roland, "Fermis as the Measure of War: Neutrons, Photons, Electrons and the Sources of Military Power," in Stephen D. Chiabotti, ed., *Tooling for War: Military Transformation in the Industrial Age* (Chicago: Imprint Publications, 1996), 173–88.

23. Ann Markusen and Joel Yudken, *Dismantling the Cold War Economy* (New York: Basic Books, 1992), 34–68; Herman O. Steckler, *The Structure and Performance of the Aerospace Industry* (Berkeley: University of California Press, 1965).

24. Linda Weiss, *America Inc.? Innovation and Enterprise in the National Security State* (Ithaca, NY: Cornell University Press, 2014), 124, 78; Vernon W. Ruttan, *Is War Necessary for Economic Growth? Military Procurement and Technology Development* (New York: Oxford University Press, 2006).

25. Rebecca U. Thorpe, *The American Warfare State: The Domestic Politics of Military Spending* (Chicago: University of Chicago Press, 2014), 78.

26. A good introduction is a 1995 Congressional Research Service report, which explained the thoughts and actions of the Department of Defense while withholding judgment on the phenomenon itself. Theodor Galdi, *Revolution in Military Affairs? Competing Concepts, Organizational Responses, Outstanding Issues*, CRS 95-1170F, Dec. 11, 1995, http://www.iwar.org.uk/rma/resources/rma/crs95-1170F.htm. For a more recent assessment, see Jeffrey Collins and Andrew Futter, eds., *Reassessing the Revolution in Military Affairs: Transformation, Evolution, and Lessons Learnt* (Basingstoke, Hampshire: Palgrave Macmillan, 2015).

27. For a skeptic's view, see Michael O'Hanlon, *Technological Change and the Future of Warfare* (Washington, DC; Brookings Institution Press, 2000).

28. White House, *National Security Strategy of the United States* (Washington, DC: White House, Jan. 1993), www.documentcloud.org /documents/3123581-1993-National-Security-Strategy.html.

29. Office of the Under Secretary of Defense (Comptroller), *National Defense Budget Estimates for FY 2015 (Green Book)*, Apr. 2014, "Table 7-5: Department of Defense Manpower," 263, www.comptroller.defense .gov/Portals/45/Documents/defbudget/fy2015/FY15_Green_Book.pdf.

30. Thomas G. Mahnken argues that service culture exerts strong influence on the choices in arms and equipment made by the different armed forces. This is certainly true, though the missions of the services probably have greater impact. Thomas G. Mahnken, *Technology and the American Way of War since 1945* (New York: Columbia University Press, 2008).

31. Quoted in Nick Turse, *The Complex: How the Military Invades Our Everyday Lives* (London: Faber & Faber, 2009), 28.

32. Armies (and the marine corps) in general pay less for weapon systems because they escape platform costs—ships for the navy and planes for the air force—though American ground forces, vis-à-vis their enemies, face exceptionally high personnel costs.

33. Richard Whittle, *Dream Machine: The Untold History of the Notorious V-22 Osprey* (New York: Simon & Schuster, 2010).

34. OMB, *Historical Tables*, table 3.1.

35. Department of Defense, Defense Manpower Data Center, accessed July 24, 2019, www.dmdc.osd.mil/appj/dwp/dwp_report.ssp.

36. Scott A. Beaulier and Joshua C. Hall, "The Impact of Political Factors on Military Base Closures," *Journal of Economic Policy Reform*, 14, no. 4 (Dec. 2011): 333–42.

37. Harold Parker, personal communication, c. 1972.

38. Rachel Weber, *Swords into Dow Shares: Governing the Decline of the Military-Industrial Complex* (Boulder, CO: Westview Press, 2001), 105–34.

39. On vertical integration, see Jacques Gansler, *The Defense Industry* (Cambridge, MA: MIT Press, 1980), 69, 103, 136–37. Gansler notes that vertical integration tends to stifle innovation.

40. In 1960, seven US companies were building nuclear submarines for the navy. By 1990, only two were left, Electric Boat and the Newport News Shipbuilding and Drydock Co. Initially, the navy had split the contract for the Seawolf-class boats between the two shipbuilders, but they could not work harmoniously. All 29 boats were assigned to Electric Boat, since Newport News also built all US carriers and was less vulnerable to the defense cutbacks. John Bilker, John Schank, Giles Smith, Fred Timson, James Chiesa, Marc Goldberg, Michael Mattock, and Malcolm MacKinnon, *The U.S. Submarine Production Base: An Analysis of Cost, Schedule, and Risk for Selected Force Structures* (Santa Monica, CA: RAND, 1994), 15.

41. John F. Schank, Cesse Ip, Frank W. Lacroix, Robert E. Murphy, Mark V. Arena, Kristy N. Kamarck, and Gordon T. Lee, *Learning from the U.S. Navy's* Ohio, Seawolf, *and* Virginia *Submarine Programs*, Vol II of *Learning from Experience* (Santa Monica, CA: RAND Corporation, 2011), 55–58, 113–14.

42. Ronald O'Rourke, *Navy Virginia (SSN-774) Class Attack Submarine Procurement: Background and Issues for Congress*, CRS Report RL32418, June 4, 2019, p. 1–3.

43. Weber, *Dow Shares*, 138.

44. Bilker, et al., *U.S. Submarine Production Base*.

45. John F. Schank, et al., "Seawolf Case Study," *Navy's Ohio, Seawolf,* *and Virginia Submarine Programs*, RAND Corporation (2011), 43–60, at 12, accessed July 15, 2019, http://www.jstor.org/stable/10.7249/j.ctt3fh0zm.12.

46. Ann Markusen, "Dismantling the Cold War Economy," *World Policy Journal* (Summer 1992): 395.

47. Michael Remez, "Senate Supporters Rally to Reject Anti-Seawolf Amendment," *The Hartford Courant*, Aug. 4, 1995, https://://www.courant .com/search/senate+supporters+rally+to+reject+anti-seawolf/100-y/.

48. Jacques S. Gansler, *Defense Conversion: Transforming the Arsenal of Democracy* (Cambridge, MA: MIT Press, 1995); Markusen and Yud-ken, *Dismantling the Cold War Economy*.

49. The view was popularized by Paul Kennedy's *The Rise and Fall of the Great Powers: Economic Change and Military Conflict from 1500 to 2000* (New York: Random House, 1987). By "burden," Kennedy meant defense spending as a percentage of GDP. See, for example, Chi Huang and Francis W. Hoole, "Military Burden and Economic Hegemonic Decline: The Case of the United States," in Alex Mintz, ed., *Political Economy of Military Spending*, 238–58; and Robert W. DeGrasse, Jr., *Military Expansion, Economic Decline: The Impact of Military Spending on U.S. Economic Performance* (Armonk, NY: M. E. Sharpe, 1983).

50. David C. Mowery and Nathan Rosenberg, *Technology and the Pursuit of Economic Growth* (Cambridge: Cambridge University Press, 1994), 123–68.

51. Michael Reich, "Military Spending and the U.S. Economy," in Steven Rosen, ed., *Testing the Theory of the Military-Industrial Complex* (Lexington, MA: D.C. Heath, 1973), 85–102.

52. Gansler, *Defense Conversion*, 54.

53. John Deutch, "Consolidation of the U.S. Defense Industrial Base," *Acquisitions Review Quarterly* 8, no. 3 (Fall 2001): 137–50.

54. Curtiss-Wright, for example, a company with roots going back to the invention of the airplane, fell by the wayside. Eugene Gholz, "The Curtiss-Wright Corporation and Cold War-Era Defense Procurement: A Challenge to Military-Industrial Complex Theory," *Journal of Cold War Studies* 2, no. 1 (Winter 2000): 35–75.

55. William D. Hartung, *Prophets of War: Lockheed Martin and the Making of the Military-Industrial Complex* (New York: Nation Books, 2011), 167–78, quote at 174. The government volunteered such help when it proposed consolidation at the last supper.

56. Augustine later donated the government-funded portion of his windfall to charity. Hartung, *Prophets of War*, 174–79.

57. Mowery and Rosenberg, *Technology and Economic Growth*, 123–68; Gansler, *Democracy's Arsenal*, 32–34.

58. Deutch, "Consolidation of the U.S. Defense Industrial Base."

59. For a stimulating sociological overview, see Fred Block and Matthew R. Keller, "Where Do Innovations Come from? Transformations in the US Economy, 1970–2006," *Socio-Economic Review*, 7 (2009), 459–83. It posits an increasing role for government.

60. Henry Kressel, "Edison's Legacy: Industrial Laboratories and Innovation," *American Affairs* 1, no. 4 (Winter 2017), www.american affairsjournal.org/2017/11/edisons-legacy-industrial-laboratories -innovation/.

61. John F. Sargent, Jr., *U.S. Research and Development Funding and Performance Fact Sheet*, CRS Report R44307, Sept. 19, 2019.

62. National Science Board, "Science and Engineering Indicators 2018," figure 4–3, Oct. 25, 2019, www.nsf.gov/statistics/2018/nsb20181 /assets/1038/figures/fig04-03.pdf.

63. AAAS, "Historical Trends in Federal R&D," Table P, By Function: Defense and Nondefense R&D, 1953–2018, accessed Oct. 25, 2019, https://www.aaas.org/programs/r-d-budget-and-policy/historical -trends-federal-rd; "Federal Funds for Research and Development, Detailed Historical Tables: Fiscal Years 1951–2002," accessed June 25, 2020, https://wayback.archive-it.org/5902/20150627201426/http:// www.nsf.gov/statistics/nsf03325/.

64. Mowery and Rosenberg, *Technology and Economic Growth*, 219–36.

65. Markusen and Yudken, *Dismantling the Cold War Economy*, 113–25. For DARPA's rationale, see Alex Roland and Philip Shiman,

Strategic Computing: DARPA and the Quest for Machine Intelligence,
1983–1993 (Cambridge, MA: MIT Press, 2002), 51–52, 90, *et passim.*

66. Sharon Weinberger, *The Imagineers of War: The Untold Story of DARPA, the Pentagon Agency that Changed the World* (New York: Knopf, 2017), 181–85; Richard D. Bingham, *Industrial Policy American Style: From Hamilton to HDTV* (Armonk, NY: M. E. Sharpe, 1998), 110–11.

67. Markusen and Joel Yudken, *Dismantling the Cold War Economy,* 33–68. Markusen and Yudken claim furthermore that DoD support for aerospace, computers, and computer electronics (ACE) was a specialized national industrial policy that privileged these military technologies over other American industries such as automobiles, petroleum, and instruments, shrinking those sectors of the economy and driving American industries and jobs overseas.

68. Tim Shorrock, *Spies for Hire: The Secret World of Intelligence Outsourcing* (New York: Simon & Schuster, 2008), 143–49; Weiss, *America Inc.?,* 65–68; Noah Shachtman, "Exclusive: Google, CIA Invest in 'Future' of Web Monitoring," Danger Room, Wired.com, July 28, 2010, www.wired.com/2010/07/exclusive-google-cia/; Paul Seabury, "Industrial Policy and National Defense," *Policy Studies Review Annual* 7 (Jan. 1, 1985): 346–566.

69. National Research Council, *Funding a Revolution: Government Support for Computing Research* (Washington, DC: National Academy Press, 1999), 129–30; Roland and Shiman, *Strategic Computing,* 92, 300, 326.

70. See chapter 3, 54–55.

71. Daniel Bell, *The Coming of Post-Industrial Society: A Venture in Social Forecasting* (New York: Basic Books, 1973); Block and Keller, "Where Do Innovations Come From?," 476; Weiss, *America Inc.?,* 4.

72. Markusen and Yudken, *Dismantling the Cold War Economy,* 85–89.

73. Smith, *Military Economics,* 124.

74. Shorrock, *Spies for Hire,* 214–21, quotes at 216, 219. Trailblazer, described by a former CIA case officer as "one of the lasting disgraces" of its era, was canceled in 2005.

75. Gansler, *Democracy's Arsenal,* 177–83.

76. Heidi M. Peters, *Department of Defense Use of Other Transaction Authority: Background, Analysis, and Issues for Congress*, CRS Report R45521, Feb. 22, 2019; Aaron Boyd, "The Scary New Contracting Model that Isn't Scary or New," *Nextgov*, Mar. 26, 2018, www.nextgov.com /it-modernization/2018/03/otas-scary-new-contracting-model-isnt -scary-or-new/146964/; Gansler, *Democracy's Arsenal*, 177–83.

77. Kate M. Manuel, "Government Contracts: Basic Legal Principles," Congressional Research Service, *In Focus*, IF10135, Feb. 23, 2015.

78. J. Ronald Fox, with contributions by David G. Allen, Thomas C. Lassman, Walton S. Moody, and Philip L. Shiman, *Defense Acquisition Reform, 1960–2009: An Elusive Goal* (Washington, DC: Center of Military History, United States Army, 2001), 190, 189. This is part of the "History of Acquisition in the Department of Defense Series."

79. Hartung, *Prophets of War*, 89–94, 100; Robert B. Semple, "Future Defense Contracts to Be Awarded in Stages," *New York Times*, July 28, 1970, www.nytimes.com/1970/07/28/archives/future-defense-contracts -to-be-awarded-in-stages-laird-adopts-fly.html.

80. William D. Hartung, *And Weapons for All* (New York: Harper-Collins, 1994); Steven Reardon, *Council of War: A History of the Joint Chiefs of Staff, 1942–1991* (Washington, DC: NDU Press, 2012,), 368.

81. Stockholm International Peace Research Institute, "TIV of arms exports from the top ten exporters, 1985–2001," accessed July 25, 2019, https://armstrade.sipri.org/html/export_toplist.php.

Chapter 9. War on Terror

1. Madeleine Albright and Bob Woodward, *Madam Secretary* (New York: Harper Perennial, 2013), 183.

2. Donald H. Rumsfeld, "Transforming the Military," *Foreign Affairs*, 81, no. 3 (May–June 2002): 2–32.

3. It is unclear why the army chose to call these artillery pieces "howitzers," which are traditionally high-trajectory, short-range guns.

4. The General Accounting Office became the Government Accountability Office in 2004.

5. James Dao, "A Lift for a Weapons System," *New York Times*, May 16, 2002.

6. Steven Lee Myers, "Pentagon Panel Recommends Scuttling Howitzer System," *New York Times*, Apr. 23, 2002.

7. James Mann, *The Rise of the Vulcans: The History of Bush's War Cabinet* (New York: Viking, 2004), 294–331.

8. National Commission on Terrorist Attacks on the United States, *The 9/11 Commission Report* (Washington, DC, 2004), http://govinfo.library.unt.edu/911/report/911Report.pdf.

9. *Uniting and Strengthening America by Providing Appropriate Tools Required to Intercept and Obstruct Terrorism Act* (USA Patriot Act) of 2001, P.L. 107-56, Oct. 26, 2001.

10. For an introduction to the National Security State, see Anna Kasten Nelson, "The Evolution of the National Security State" in Andrew J. Bacevich, ed., *The Long War: A New History of National Security Policy since World War II* (New York: Columbia University Press, 2007), 265–301.

11. Jacques Gansler concluded, however, that the War on Terrorism was fought with a pre-9/11 arsenal. Jacques S. Gansler, *Democracy's Arsenal: Creating a Twenty-First Century Defense Industry* (Cambridge, MA: MIT Press, 2011), 339.

12. Mann, *Rise of the Vulcans,* 300–302.

13. Global Security, "U.S. Casualties in Iraq," accessed Nov. 7, 2019, www.globalsecuirty.org/military/ops/iraq_casualties.htm; "Iraq Body Count," accessed Nov. 7, 2019, www.iraqbodycount.org.

14. Mann, *Rise of the Vulcans,* 227–28, 273.

15. Government Accounting Office, 2009, quoted in Jacques S. Gansler, William Lucyshyn, and William Varettoni, *Acquisition of Mine-Resistant, Ambush-Protected (MRAP) Vehicles: A Case Study*, School of Public Policy, University of Maryland: NPS Acquisition Research Symposium, May 12, 2010, p. 5.

16. John Pike, "Mine Resistant Ambush Protected (MRAP) Vehicle Program," accessed Aug. 5, 2019, https://www.globalsecurity.org/military/systems/ground/mrap.htm.

17. Thomas J. Stafford, "The MRAP Vehicle: The New Icon of Operation Iraqi Freedom," *Infantry* (Nov.–Dec. 2007), 16.

18. Sharon Weinberger, *The Imagineers of War: The Untold Story of DARPA, the Pentagon Agency that Changed the World* (New York: Knopf, 2017), 210.

19. James E. Hasik, *Securing the MRAP: Lessons Learned in Marketing and Military Procurement* (College Station: Texas A&M University Press, 2021).

20. Michael E. Bulkley and Gregory C. Davis, "The Study of the Rapid Acquisition Mine Resistant Ambush Protected (MRAP) Vehicles Program and Its Impact on the Warfighter," Naval Postgraduate School, Monterey California, June 2013, v. As in the crash program to develop atomic weapons in World War II, multiple possibilities were pursued simultaneously.

21. Gansler, *Acquisition of MRAP Vehicles*, 4.

22. Stafford, "The MRAP Vehicle," 16–19.

23. Furthermore, the threat should have come as no surprise. In the Chinese civil war, Mao Zedong had taught the world how insurgents could challenge industrialized armies by ambushing their vehicles on the roads to which they were bound by their own logistic burdens. The United States had to relearn the lesson in Vietnam. The hubris of the RMA contributed to American forces being caught flatfooted by an enemy asymmetric technological innovation.

24. Even expedited delivery of the vehicles to the war zone posed significant problems. It cost about $135,000 to fly each vehicle to the Middle East. Adam Hebert, "For Quick Delivery of Mine-Resistant Vehicles, U.S. Military Relies on Airlift," *World Politics Review*, Apr. 7, 2008, www.worldpoliticsreview.com/articles/1902/for-quick-delivery-of-mine-resistant-vehicles-u-s-military-relies-on-airlift.

25. Bulkley and Davis, "MRAP Vehicles Program," 58–59.

26. Pike, "MRAP Vehicle Program."

27. Chris Rohlfs and Ryan Sullivan, "The MRAP Boondoggle," *Foreign Affairs* Snapshot, July 26, 2012, www.foreign affairs.com/articles/137800/chris-rohlfs-and-ryan-sullivan/the-mrap-boondoggle; Jason

Shell, "How the IED Won: Dispelling the Myth of Tactical Success and Innovation, War on the Rocks," May 1, 2017, https://warontherocks .com/2017/05/how-the-ied-won-dispelling-the-myth-of-tactical-success -and-innovation/.

28. Ann Daugherty Miles, *Intelligence Spending: In Brief,* CRS Report 44381, Feb. 16, 2016, p. 8.

29. The major categories of technical means are signals intelligence (sigint), geospatial intelligence (geoint), and measurement and signature intelligence (masint). Anne Daugherty Miles, *Intelligence Community Programs, Management, and Enduring Issues,* Congressional Research Service Report R44681, Nov. 8, 2016, pp. 12–13.

30. John Pike, the highly respected director of GlobalSecurity.org, has reported that much of the intelligence budget was listed as "other procurement aircraft" in the air force's budget. The significant point is that US spending on intelligence has long been part of US military spending.

31. The "topline" budget is a colloquial term for the aggregate total for that funding category, what the business community would call the bottom line. Michael E. DeVine, "Intelligence Community Spending: Trends and Issues," CRS Report 44381, June 18, 2018, p. 1.

32. There are some other expenditures for intelligence, such as the Homeland Security Intelligence Program, that are not covered by either the NIP or MIP. DeVine, "Intelligence Community Spending," 1–2.

33. See chapter 8, 125–26.

34. Wendy Molzahn refers to In-Q-Tel as a "venture catalyst" firm. Wendy Molzahn, "The CIA's In-Q-Tel Model: Its Applicability," *Acquisition Review Quarterly* 10, no. 1 (Winter 2003): 46–61.

35. Linda Weiss, *America Inc.? Innovation and Enterprise in the National Security State* (Ithaca, NY: Cornell University Press, 2014), 68.

36. Quoted in Tim Shorrock, *Spies for Hire: The Secret World of Intelligence Outsourcing* (New York: Simon & Schuster, 2008), 144–45.

37. Weiss, *America Inc.?,* 51–74. The limits of the US Cold War R&D model had already been revealed in a brilliant and insightful study, *Beyond Spinoff.* This jointly authored study by five of America's most astute students of government organization also endorsed dual-use technologies and other innovations that would come to pass in the twenty-first century. John Alic, Lewis M. Branscomb, Harvey Brooks, Ashton B. Carter, and Gerald L. Epstein, *Beyond Spinoff: Military Commercial Technologies in a Changing World* (Boston: Harvard Business School Press, 1992).

38. Alic, et al., *Beyond Spinoff*; David C. Mowery and Nathan Rosenberg, *Technology and the Pursuit of Economic Growth* (Cambridge: Cambridge University Press, 1994),137–50, 199–200; Richard J. Samuels, *Rich Nation, Strong Army: National Security and the Technological Transformation of Japan* (Ithaca, NY: Cornell University Press, 1994), 18–31.

39. "CACI International Inc. History," *Funding Universe*, accessed Aug. 26, 2019, http://www.fundinguniverse.com/company-histories/caci-international-inc-history/.

40. Shorrock, *Spies for Hire*, 229.

41. Scott Shane, "John Michael McConnell, a Member of the Club," *New York Times*, Jan. 5, 2007, https://www.nytimes.com/2007/01/05/washington/05mcconnell.html; Booz Allen Hamilton, "Meet Mike," accessed Aug. 26, 2019, www.boozallen.com/d/bio/leadership/mike-mcconnell.html.

42. Shorrock, *Spies for Hire*, 156–81.

43. The success in making the United States a "hard target" resulted from a combination of traveler screening, intelligence gathering, and improved inter-agency information-sharing. See Peter Bergen and David Sterman, "Jihadist Terrorism: 17 Years after 9/11," *New America*, Sept. 10, 2018, https://www.newamerica.org/international-security/reports/jihadist-terrorism-17-years-after-911/.

44. CSRA, created on Nov. 30, 2015, out of the merger of two other corporate contractors for the federal government, existed for just 28 months before disappearing into General Dynamics.

45. Tim Shorrock, "5 Corporations Now Dominate Our Privatized Intelligence Industry," *The Nation*, Sept. 8, 2016, www.thenation.com/article/five-corporations-now-dominate-our-privtized intelligence-industry/; Ross Wilkers, "General Dynamics Takes Top Spot with CSRA

Addition" *Washington Technology,* June 1, 2018, www.washing
tontechnology.com/articles/2018/06/03/gd-top-100-profile.aspx.

46. See, for example, John A. Alic, *Trillions for Military Technology:
How the Pentagon Innovates and Why It Costs So Much* (New York:
Palgrave Macmillan, 2007), 196–200;

47. Carl Conetta, "Trillions to Burn: A Quick Guide to the Surge in
Pentagon Spending," Project on Defense Alternatives, www.comw.org
/pda/1002BudgetSurge.html, accessed June 27, 2019. Of course, this
gross number becomes less meaningful over time, as US military force
becomes increasingly automated and many military functions are
outsourced.

48. P. W. Singer, *Corporate Warriors: The Rise of the Privatized
Military Industry* (Ithaca: Cornell University Press, 2003), 136–46;
Pratap Chatterjee, *Halliburton's Army: How a Well-Connected Texas Oil
Company Revolutionized the Way America Makes War* (New York:
Nation Books, 2009), 51–65; Dan Brioly, *The Halliburton Agenda: The
Politics of Oil and* Money (Hoboken, NJ: Wiley, 2004), 181–237; Valerie
Bailey Grasso, "Defense Logistical Support Contracts in Iraq and Afghan-
istan: Issues for Congress," CRS, Report RL33834, Apr. 28, 2010.

49. Singer, *Corporate Warriors,* 4–5.

50. Singer, *Corporate Warriors,* 120.

51. Singer, *Corporate Warriors,* 152–58.

52. Singer, *Corporate Warriors,* 151–90, quotes at 155, 171.

53. US Army Colonel Bruce Grant, quoted in Singer, *Corporate
Warriors,* 204.

54. Singer, *Corporate Warriors,* 222, 236. See also, David Isenberg, "It's
Déjà Vu for Dyncorp All Over Again," May 25, 2011, https://huffpost.com
/entry/its-dj-vu-for-dyncorp-all_b_792394.

55. Singer, *Corporate Warriors,* 239.

56. Paul C. Light, "The True Size of Government: Tracking Washington's
Blended Workforce, 1984–2015," The Volker Alliance, Issue Papers, Oct. 5,
2017, https://www.volckeralliance.org/publications/true-size-government.

57. "Total Force – The Cost Component FY01–10," accessed Aug. 27,
2019, http://www.pogoarchives.org/m/co/enclosure1-20121017.pdf.

58. Paul C. Light, "Fact Sheet on the True Size of Government," accessed Aug. 27, 2019, www.wagner.nyu.edu/files/faculty/publications/lightFact TrueSize.pdf.

59. Dana Priest and William M Arkin, *Top Secret America: The Rise of the New American Security State* (New York: Little, Brown, 2011), 187–88.

60. Light, *The Government-Industrial Complex,* 44. Overall, however, civilian agencies spent 78% of their budgets on service contracts and grants in 2015, while the DoD spent 53% of their slightly larger total (p. 13). In 2011, 60% of DoD contracts were for services. Gansler, *Democracy's Arsenal,* 298.

61. Noah Coburn, *Under Contract: The Invisible Workers of America's Global War* (Stanford, CA: Stanford University Press, 2018), 16.

62. Blackwater USA (1997–2002) has changed its name, leadership, and ownership several times, from Blackwater Security Consulting (2002–2007) to Blackwater Worldwide (2007–2009), Xe Services (2009–2010), Academi (2010–2014), and Constellis Holdings (since 2014).

63. Peter W. Singer, "The Dark Truth about Blackwater," *Brookings,* Oct. 2, 2007, https://brookings.edu/articles/the-dark-truth-about -blackwater/.

64. Congress of the United States, Congressional Budget Office, *Contractors' Support of U.S. Operations in Iraq* (Washington, DC, Aug. 2008), 14. It is important to note that the government did not have direct data on the number of contractor personnel then supporting the military and other government agencies abroad. Rather it was using procurement data from the Federal Procurement Data System—Next Generation (FPDS) to estimate personnel numbers based on the amounts of personal services contracts.

65. James McCartney and Molly Sinclair McCartney, *America's War Machine: Vested Interests, Endless Conflicts* (New York: St. Martin's, 2015), 30.

66. Robert W. Wood, "Independent Contractor v. Employee and Blackwater," *Montana Law Review* 70, no. 1 (Winter 2009): 96.

67. See the "Taguba Report," "Article 15–6 Investigation of the 800th

Military Police Brigade," June 4, 2004, www.fas.org/irp/agency/dod
/taguba.pdf.

68. One "CIA officer who oversaw the agency's program at the Abu
Ghraib prison" was later accused of supervising "an unofficial program in
which the CIA imprisoned and interrogated men without entering their
names in the Army's books." This so-called "ghosting program" resulted in
the single death reported at Abu Ghraib. It is unclear if any of the
contractors, usually described in the press as "military contractors," were
actually employed directly by the CIA. The accused agent reportedly
resigned from the CIA shortly after the death, received a letter of repri-
mand, and later "rejoined the intelligence community as a contractor."
Apparently, no one was ever charged with what the military called a
homicide. "Ex-CIA Officer's Role in Abu Ghraib Death Probed," *CBS
News*, July 13, 2011, www.cbsnews.com/news/ex-cia-officers-role
-in-abu-ghraib-death-probed/.

69. P. W. Singer, "Outsourcing War," *Foreign Affairs*, 84, no. 2 (Mar.–
Apr. 2005), 119–32.

70. Linda Weiss reports, for example, that Lockheed Martin, the
world's largest defense contractor, in addition to running "interrogation
programs and conducting security operations, . . . sorts mail and calcu-
lates taxes, cuts social security checks, runs space flights, and monitors air
traffic." Weiss, *America Inc.?*, 152.

71. Edward Jay Epstein, *How America Lost Its Secrets: Edward
Snowden: The Man and the Theft* (New York: Knopf, 2017).

Chapter 10. A Peer Rival?

1. John A. Alic, *Trillions for Military Technology: How the Pentagon
Innovates and Why It Costs So Much* (New York: Palgrave Macmillan,
2007), 191–96.

2. By the summer of 2007, Tenet had made almost $3 million directing
and advising IIC contractors. Tim Shorrock, *Spies for Hire: The Secret
World of Intelligence Outsourcing* (New York: Simon & Schuster, 2008),
28–29.

3. Congressional Budget Office, *Funding for Overseas Contingency Operations and Its Impact on Defense Spending*, CBO Report 54219, Oct. 23, 2018, www.cbo.gov/system/files/2018-10/54219-oco_spending.pdf. It should be noted that the same dodge has been used to supplement funding for the "foreign affairs agencies." See Brenda W. McGarry and Emily M. Morgenstern, *Overseas Contingency Operations Funding: Background and Status*, CRS Report 44519, Sept. 6, 2019.

4. Office of the Under Secretary of Defense (Comptroller), *National Defense Budget Estimates for FY 2019*, Apr. 2018, table 2–1, www.COMP TROLLER.defense.gov/Portals/45/Documents/defbudget/fy2019 /FY19_Green_Book.pdf.

5. White House, Office of Management and Budget, Historical Tables, table 3.1, accessed Nov. 20, 2019, www.whitehouse.gov/omb/historical -tables/.

6. Jim Wolfe, "Powell: 'I'm Running Out of Demons,'" *Army Times*, Apr. 15, 1991, 4. Powell had learned in Vietnam to beware of committing US troops abroad. He and Secretary of Defense Caspar Weinberger introduced into public discourse two versions of what came to be called the Weinberger-Powell Doctrine, a set of preconditions to be met before the United States committed American ground forces to a major foreign expedition. Jeffrey Record, "Back to the Weinberger-Powell Doctrine?," *Strategic Studies Quarterly* 1, no. 1 (Fall 2007): 79–95.

7. "Breaking: Gates Stuns Navy League With Fleet Proposals," *New Wars*, May 3, 2010, https://newwars.wordpress.com/2010/05 /03/breakinggates-stuns-navy-league-with-fleet-proposals/.

8. Lockheed CEO Norman Augustine collected his tongue-in-cheek precepts into *Augustine's Laws*, a popular book that went through six editions. Law number 16 said that in 2054, the defense budget would be able to afford only one new airplane.

9. Bob Woodward, *Obama's Wars* (New York: Simon and Schuster, 2011), 372.

10. "Understanding Sequester: An Update for 2018," Mar. 12, 2018, https://budget.house.gov/sites/democrats.budget.house.gov/files/doc uments/sequester%20update%20post-BBA%20FINAL.pdf.

11. Grant A. Driessen and Megan S. Lynch, *The Budget Control Act: Frequently Asked Questions*, CRS Report R44874, Jan. 22, 2017–Feb. 23, 2018, p. 3.

12. Paul Kane, "Congress Is Losing One of the Only Incentives It Had to Address the Deficit," *Washington Post*, July 27, 2019, www.washing tonpost.com/powerpost/congress-is-losing-one-of-the-only-incentives -it-had-to-address-the-deficit/2019/07/27/752fae9c-afca-11e9-bc5c -e73b603e7f38_story.html.

13. Congressional Budget Office, "Funding for Overseas Contingency Operations and Its Impact on Defense Spending," Report 54219, Oct. 23, 2018, www.cbo.gov/publication/54219/.

14. Linda Weiss, *America Inc.? Innovation and Enterprise in the National Security State* (Ithaca: Cornell University Press, 2014), 69, 73, 156; "National Technology Alliance Awards, GUARD Program," *Government* Technology, June 14, 2005, https://www.govtech.com/policy-man agement/National-Technology-Alliance-Awards-GUARD-Program .html.

15. "Venture Acceleration Fund," July 14, 2017, www.federallabs.org /news/venture-acceleration-fund.

16. Weiss, *America Inc.?*, 126–27.

17. For examples of innovative reforms advanced during the George H. W. Bush administration, see US Congress, Office of Technology Assessment, *Redesigning Defense: Planning the Transition to the Future U.S. Defense Industrial Base*, OTA-ISC-500 (Washington, DC: Government Printing Office, 1991); *Idem., After the Cold War: Living with Lower Defense Spending* (Washington, DC: Congress of the United States, Office of Technology Assessment, February 1992); John A. Alic, Lewis M. Branscomb, Harvey Brooks, Ashton B. Carter, and Gerald L. Epstein, *Beyond Spinoff: Military and Commercial Technologies in a Changing World* (Boston: Harvard Business School Press, 1992).

18. Scott Hubinger, "Can the F-35 Lightning II Joint Strike Fighter Avoid the Fate of the F-22 Raptor?," *Joint Force Quarterly 94* (3rd Quarter 2019), 44–52, at 50.

19. P. Kathie Sowell, "The C⁴ISR Architecture Framework: History,

Status, and Plans for Evolution," accessed Nov. 2, 2019, www.apps.dtic
.mil/dtic/tr/fulltext/u2/a456187.pdf; John Ferris, "Netcentric Warfare,
C⁴ISR and Information Operations: Towards a Revolution in Military
Intelligence," *Intelligence and National Security*, 19, no. 2 (2004),
199–225, accessed Nov. 2, 2019, www.dori.org/10.1080/0268452042000
302967. C⁴ISR also has roots in the theories of John Boyd; see Tim Grant
and Bas Kooter, "Comparing OODA & Other Models as Operational View
C² Architecture," *The Future of C²: 10th International Command and
Control Research and Technology Symposium*, Royal Netherlands
Military Academy, June 13–16, 2005, www.pdfs.semanticscholar.org
/b190/1ae2dd74bd5ff96dc2c350c13e4ca4091de9.pdf.

20. Past abuses, it should be noted, failed to disappear. In 2018,
Senator Chuck Grassley (R, IA) still found himself asking how the air
force could possibly justify paying $1,280 for self-warming coffee cups.
"Grassley Presses Air Force for Answers on $1,280 Cup," accessed
July 19, 2019, https://www.grassley.senate.gov/news/news-releases
/grassley-presses-air-force-answers-1280-cup.

21. See Committee on Innovations in Computing and Communica-
tions, National Research Council, *Funding a Revolution: Government
Support for Computing Research* (Washington, DC: National Academy
Press, 1999); Alic, *Trillions for Military Technology*, 186–88.

22. William D. Hartung, *Prophets of War: Lockheed Martin and the
Making of the Military-Industrial Complex* (New York: Nation Books,
2011), 89–132.

23. This familiar trope in the "national security" literature harkens
back, of course, to Harold Lasswell's "garrison state," which so captured
Dwight Eisenhower's imagination and fueled his concern about a
"military-industrial complex."

24. Weiss, *America Inc.?*, 181. This assertion will be examined more
closely in the "Conclusion." For now, suffice it to say that the implication of
the term "national security state" seems to exaggerate the dominance of
security issues in public consciousness, federal spending, and political
influence.

25. Alex Roland and Philip Shiman, *Strategic Computing: DARPA and* *the Quest for Machine Intelligence, 1983–1993* (Cambridge, MA: MIT Press, 2002).

26. Robert Higgs, "U.S. Military Spending in the Cold War Era: Opportunity Costs, Foreign Crises, and Domestic Constraints," Cato Institute Policy Analysis No. 114," Nov. 10, 1988, www.object.cat.org/pubs/pas/pa114.pdf.

27. F/A designates a fighter/attack aircraft. This durable plane went through multiple modifications, just now going into retirement in its 18E/F Super Hornet version in anticipation of the F-35.

28. Christopher J. Niemi, "The F-22 Acquisition Program: Consequences for the U.S. Air Force's Fighter Fleet," *Air & Space Power Journal* (Nov.–Dec. 2012), 53–82, quote at 69.

29. Nick Turse, *The Complex: How the Military Invades Our Everyday Lives* (London: Faber & Faber, 2009), 28.

30. The air force began planning for the F-35 in 1992, just one year after initiating the F-22 program.

31. Mark Thompson, "The Most Expensive Weapon Ever Built," *TIME Magazine* 181, no. 7 (Feb. 25, 2013), www.content.time.com/time/magazine/article/0,9171,2136312,00.html.

32. T. Schóber and P. Puliš, "F-35 - Win or Loss for the USA and their Partners?," *Advances in Military Technology* 10, no. 2 (Dec. 2015), 84.

33. Swadesh Rana, "Problems of U.S. European Co-Production in Arms," *Strategic Analysis* 3, no. 7 (1979): 272–76; I. Megens, "Problems of Military Production Co-ordination," in B. Heuser and R. O'Neill, eds., *Securing Peace in Europe, 1945–62* (London: Palgrave Macmillan, 1992), 279–92.

34. Srdjan Vucetic and Kim Richard Dossal, "The International Politics of the F-35 Joint Strike Fighter," *International Journal* (Winter 2012–13): 5.

35. John Mintz, "The Fighters on Which They Bet the Farm," *Washington Post*, Nov. 8, 1996, C11. This article quotes Lockheed Martin's CEO, Norman Augustine, as saying "You're betting your company's life." Some critics asserted that the air force had earlier "bet the farm" on Lockheed's F-104, with disastrous results.

36. Michael P. Hughes, "The F-35—Not the Super Fighter We Were Promised: A Decade behind Schedule, the Most Expensive Program in World History Is Failing," *The Washington Spectator*, Aug. 14, 2017, www .washingtonspectator.org/hughes/f-35.

37. Michael P. Hughes, "Lockheed Martin and the Controversial F-35," *Journal of Business Case Studies* 11, no. 1 (First Quarter 2015): 1–14.

38. Quoted in Valerie Insinna, "Inside America's Dysfunctional Trillion-Dollar Fighter-Jet Program," *New York Times Magazine*, Aug. 21, 2019, www.nytimes.com/2019/08/21/magazine/ 35-joint-strike -fighter-program.html, 4.

39. Insinna, "Trillion Dollar Program," 10.

40. Insinna, "Trillion Dollar Program," 3.

41. Hughes, "Lockheed Martin," 10.

42. Insinna, "Trillion Dollar Program," 5; Thompson, "Most Expensive Weapon," 4. Raphaël Zaffran and Nicolas Erwes use path dependence, "lock-in" and the "sunk cost trap or 'paradox'" to explain in social science terms what Thomas P. Hughes called "momentum." Raphaël Zaffran and Nicolas Erwes, "Beyond the Point of No Return? Allied Defence Procurement, the 'China Threat,' and the case of the F-35 Joint Strike Fighter," *Asian Journal of Public Affairs*, 8, no. 1 (2015): 69–71; Thomas P. Hughes, "Technological Momentum," in Merrit Roe Smith and Leo Marx, eds., *Does Technology Drive History? The Dilemma of Technological Determinism* (Cambridge, MA: MIT Press, 1994), 101–13.

43. Insinna, "Trillion Dollar Program," 7.

44. Germany, for example, ruled out the F-35 as a replacement for its Tornado fighter. Sebastian Sprenger, "Germany Officially Knocks F-35 Out of Competition to Replace Tornado," *DefenseNews*, Jan. 31, 2019, https://www.defensenews.com/global/europe/2019/01/31/germany-offi cially-knocks-f-35-out-of-competition-to-replace-tornado/. As early as 2013, Canada and Australia reported reconsidering their planned purchases. Thompson, "Most Expensive Weapon," 5.

45. Idrees Ali and Phil Stewart, "U.S. Removing Turkey from F-35 Program after Its Missile Defense Purchase," *Reuters World News*, July 17, 2019, https://www.reuters.com/article/us-usa-turkey-security-f35

/u-s-removing-turkey-from-f-35-program-after-its-russian-missile
-defense-purchase-idUSKCN1UC2GL; Schóber and Puliš, "F-35—Win or
Loss?," 89.

46. Insinna, "Trillion-Dollar Program," 7. Mark Thompson says to
"think of it as a flying Swiss army knife." Thompson, "Most Expensive
Weapon," 3.

47. Quoted in Hughes, "Not the Super Fighter We Were Promised."

48. Tom Burbage, quoted in Insinna, "Trillion Dollar Program," 3.

49. Carlo Kopp, "Lockheed Martin's F-35 Joint Strike Fighter,"
Australian Aviation (Apr.–May 2002): 24–32, www.http://ausairpower
.net/jsf-analysis-2002.html.

50. Loren Thompson, "Super-Weapon: Why Have F-35 Fighter Costs
Increased," *Forbes*, Oct. 15, 2012, https://www.forbes.com/sites/loren
thompson/2012/10/15/super-weapon-why-have-f-35-fighter-costs
-increased/#204521e93ee1. Thompson attributes cost escalation to
government demands for changes to the contract more than to Lockheed
malpractice.

51. For a longer list, see Todd S. Sechser, Neil Narang, and Cait-
lin Talmadge, "Emerging Technologies and Strategic Stability in Peacetime, Crisis, and War," *Journal of Strategic Studies* 42, no. 6 (2019):
727–35.

52. Good introductions are Sarah Kreps, *Drones: What Everyone Needs
to Know* (New York: Oxford University Press, 2016); Michael J. Boyle, *The
Drone Age: How Drone Technology Will Change War and Peace* (New
York: Oxford University Press, 2020).

53. Peter Bergen, Melissa Salyk-Virk, and David Sterman, "World of
Drones," New America, *International Security*, Nov. 22, 2019, www
.newamerica.org/international-security/reports/world-drones/; Peter W.
Singer, "Do Drones Undermine Democracy," *New York Times*, Jan. 21,
2012, www.law/upenn/institutes/cerl/conferences/targetedkilling
/papers/SingerDoDronesUndermineDemocracy.pdf, 280–90.

54. Boyle, *The Drone Age*, 272–92.

55. Thomas Rid, *Cyber War Will Not Take Place* (New York: Oxford
University Press, 2013). In Rid's view, the alleged 2009 US-Israeli

"Stuxnet" cyberattack on Iran's nuclear enrichment centrifuges, and other such reported strikes, are sabotage, not war.

56. Shane Harris, *@War: The Rise of the Military-Internet Complex* (Boston: Houghton Mifflin, 2014), xix.

57. The White House, Office of Management and Budget, "President's Budget" for 2019, pp. 273–75.

58. Harris, *@War*, xix, 219.

59. P. W. Singer, *Wired for War: The Robotics Revolution and Conflict in the 21st Century* (New York: Penguin, 2009).

60. Singer, *Wired for War*, 124–34.

61. Christian Brose, "The New Revolution in Military Affairs," *Foreign Affairs* (May/June 2019), 122–34; James Hasik, "Beyond the Third Offset: Matching Plans for Innovation to a Theory of Victory," *Joint Forces Quarterly*, 91 (4th Quarter, 2018): 14–21.

Conclusion

1. John Lewis Gaddis, *The Long Peace: Inquiries into the History of the Cold War* (New York: Oxford University Press, 1987). It is possible to treat the Korean War as a great-power war, but Gaddis and other students of the Cold War accept the American euphemisms of "limited war" or "police action."

2. I know of no book-length defense of the MIC since John Stanley Baumgartner, *The Lonely Warriors; Case for the Military-Industrial Complex* (Los Angeles: Nash Pub., 1970).

3. Even Powell could not match Marshall's record of moving on to become secretary of defense. Only Ulysses S. Grant and Dwight D. Eisenhower outshone their transitions from military to civilian sectors of national leadership.

4. Kathleen McInnes, "Goldwater-Nichols at 30: Defense Reform and Issues for Congress," Congressional Research Service Report R44474, June 2, 2016.

5. General McMaster remained on active duty while serving as national security advisor to President Donald Trump in 2017 and 2018.

6. Benjamin O. Fordham, "Paying for Global Power: Assessing the Costs and Benefits of Postwar U.S. Military Spending," in Andrew J. Bacevich, ed., *The Long War: A New History of National Security Policy since World War II* (New York: Columbia University Press, 2007), 371–404.

7. Mark Wilson, "The Military-Industrial Complex," in David Kieran and Edwin A. Martini, eds., *At War: The Military and American Culture in the Twentieth Century and Beyond* (New Brunswick, NJ: Rutgers University Press, 2018), 67–86.

8. NASA, for example, an agency born in the Cold War and derived in large measure from military roots, has experienced cost escalation in virtually all of its major programs. An apocryphal tale from NASA's early history was that it escaped this pattern only in the Apollo program, because NASA Administrator James E. Webb unilaterally doubled the agency's collective estimate of the cost of the program when he was presenting the projection to Congress, sensing that he had more leverage then than he would at any time in the future.

9. For focused case studies of five programs, see Theo Farrell, *Weapons without a Cause: The Politics of Weapons Acquisition in the United States* (New York: St. Martin's Press, 1997). An abridgement appears in Theo Farrell, "Weapons without a Cause: Buying Stealth Bombers the American Way," *Contemporary Security Policy* 14, no. 2 (1993): 115–50.

10. United States Government Accountability Office [GAO], "Weapon Systems Annual Assessment," Report to Congress, GAO-19-336SP (Washington, DC: Government Printing Office, May 2019), 15.

11. GAO, "Weapon Systems Annual Assessment," 3.

12. Paul C. Light, *The Government-Industrial Complex: The True Size of the Federal Government, 1984–2018* (New York: Oxford University Press, 2019), 27–29.

13. Mark Skidmore and Catherine Austin Fitts, "Summary Report on 'Unsupported Journal Voucher Adjustments' in Financial Statements of the Office of the Inspector General for the Department of Defense and the Department of Housing and Urban Development," accessed Jan. 14, 2019, https://missingmoney.solari.com/wp-content/uploads/2018/08

/Unsupported_Adjustments_Report_Final_4.pdf. Losses at the DoD greatly outweighed those at HUD.

14. See Government Accountability Office [GAO], *High Risk Series,* Report to Congressional Committees, GAO-19-157SP (March 2019), 143–69.

15. Since 2001, the Department of Defense has spent more on contract than noncontract outlays. Jesse Ellman, David Morrow, and Gregory Sanders, "U.S. Department of Defense Contract Spending and the Supporting Defense Industrial Base," A Report to the CSIS Defense Industrial Initiatives Group (Washington, DC: Center for Strategic and International Studies, September 2012), 5.

16. Light, *Government-Industrial Complex,* 40.

17. Donna Martin, "Defense Procurement Information Papers: Campaign '84," Project on Military Procurement (Washington, DC: Fund for Constitutional Government, Aug. 1984).

18. Under Secretary of Defense for Acquisition, Technology and Logistics, "Report to Congress on Contracting Fraud," Oct. 2011.

19. Stephen Rodriguez, "Top 10 Failed Defense Programs of the RMA Era," *War on the Rocks,* Dec. 2, 2014, https://warontherocks.com /2014/12/top-10-failed-defense-programs-of-the-rma-era/.

20. United States Government Accountability Office, *Weapon Systems Annual Assessment: Limited Use of Knowledge-Based Practices Continues to Undercut DOD's Investments,* GAO-19–336SP. Washington, DC: May 2019.

21. Federation of American Scientists, "Defense Contracting Fraud: A Persistent Problem," accessed Aug. 2, 2019, www.fas.org/blogs/secrecy /2019/05/defense-contracting-fraud/.

22. Project on Government Oversight, "Federal Contractor Misconduct Database," accessed Jan. 20, 2020, https://www.contractormisconduct .org/.

23. Jerry Kammer, Dean Calbreath, and George E. Condon, Jr., *The Wrong Stuff: The Extraordinary Saga of Randy "Duke" Cunningham, the Most Corrupt Congressman Ever Caught* (New York: PublicAffairs, 2007).

24. Light, *The Government-Industrial Complex,* 154–55. After the first

Defense Base Realignment and Closure Commission of 1988, five more commissions operated from 1990 to 2005 under the Defense Base Realignment and Closure Act of 1990 and subsequent statutes.

25. An important body of scholarship, however, finds little direct correlation between location of defense contractors and the voting behavior of members of Congress and senators representing those districts and states. See, for example, Rebecca U. Thorpe, *The American Warfare State: The Domestic Politics of Military Spending* (Chicago: University of Chicago Press, 2014); Barry S. Rundquist and Thomas M. Casey, *Congress and Defense Spending: The Distributive Politics of Military Procurement* (Norman: University of Oklahoma Press, 2002); Alex Mintz, "Guns versus Butter: A Disaggregated Analysis" in Alex Mintz, ed., *The Political Economy of Military Spending in the United States* (London: Rutledge, 1992), 185–95.

26. Frank Newport, "Americans' Confidence in Institutions Edges Up," *Gallup*, June 26, 2017, www.news.gallup.com/poll/212840/americans -confidence-institutions-edges.aspx; Niall McCarthy, "The Institutions Americans Trust Most and Least in 2018," *Forbes*, June 29, 2018, https:// www.forbes.com/sites/niallmccarthy/2018/06/29/the-institutions -americans-trust-most-and-least-in-2018-infographic/#73c60fe42fc8.

27. Leo Shane III, "Veterans in the 116th Congress, by the numbers," *Military Times*, Nov. 20, 2018), https://www.militarytimes.com/news /pentagon-congress/2018/11/21/veterans-in-the-116th-congress-by-the- numbers/.

28. Article 1, Section 6, Clause 2 of the Constitution says in part that "no Person holding any Office under the United States, shall be a Member of either House during his Continuance in Office."

29. Peter D. Feaver, *Armed Servants: Agency, Oversight, and Civil- Military Relations* (Cambridge, MA; Harvard University Press, 2003), 55–74, 181, 210–14, *et passim*. "Shirking" is part of Feaver's "agency theory" of civil-military relations, which aims to replace the path- breaking conceptualization of Samuel P. Huntington, *The Soldier and the State: The Theory and Politics of Civil-Military Relations* (Cambridge, MA: Belknap Press of Harvard University Press, 1957).

30. James Fallows, "The Tragedy of the American Military," *The Atlantic*, Jan.–Feb. 2015, 79.

31. Fred Block and Matthew R. Keller, "Where Do Innovations Come From? Transformations in the US Economy, 1970–2006," *Socio-Economic Review* 7 (2009): 459–83.

32. United States, White House, Office of Management and Budget [OMB], *Historical Tables*, table 3.1.

33. Fareed Zakaria, "Defense Spending is America's Cancerous Bipartisan Consensus," *Washington Post*, July 18, 2019, www.washing tonpost.com/opinions/defense-spending-is-americas-cancerous-biparti san-consensus/2019/07/18/783a9e1a-a978-11e9-9214-246e594de5d5 _story.html; Jessica T. Mathews, "America's Indefensible Defense Budget," *The New York Review of Books,* July 18, 2019, www.nybooks.com/articles /2019/07/18/americas-indefensible-defense-budget/.

34. SIPRI *Military Expenditure* Database, accessed on Dec. 3, 2019, https://www.sipri.org/databases/milex.

35. OMB, *Historical Tables*, table 3.1.

36. William D. Hartung and Mandy Smithburger, "America's Defense Budget Is Bigger Than You Think," *The Nation*, May 7, 2019, www .thenation.com/article/tom-dispatch-america-defense-budget-bigger -than-you-think; Robert Higgs, "The Trillion-Dollar Defense Budget Is Already Here," *The Independent Institute*, Mar. 15, 2007, www.indepen dent.org/news/article.asp?id=1941. Ever since World War II, the real cost of "national security" has significantly exceeded "military spending," but for comparisons over time and across national boundaries the latter category must suffice. Furthermore, during the Cold War and beyond many of the nonmilitary costs of national security (intelligence, nuclear weapons, covert operations, etc.) were hidden within the defense budget, while others such as Veterans Affairs were hiding in plain sight.

37. Dwight D. Eisenhower, "The Chance for Peace," delivered before the American Society of Newspaper Editors, Apr. 16, 1953, https://www .eisenhowerlibrary.gov/sites/default/files/file/chance_for_peace.pdf.

38. Vernon W. Ruttan, *Is War Necessary for Economic Growth?*

Military Procurement and Technology Development (New York: Oxford University Press, 2006), 9.

39. Weiss, *America Inc.?*, 124, 143.

40. J. Paul Dunne and Mehmet Uye, "Defense Spending and Development," in T. H. Tan, ed., *The Global Arms Trade: A Handbook* (London: Rutledge, 2010), 293–305; Todd Sandler and Keith Hartley, *The Economics of Defense* (Cambridge: Cambridge University Press, 1995), 215–20.

41. See, for example, J. Paul Dunne and Derek Braddon, "Economic Impact of Military R&D," School of Economics, Bristol Business School, University of the West of England, Bristol, *Report* (June 2008). To these must be added concerns that defense industries suffer more legal and ethical lapses than their counterparts in the private sector because their government overseers too often look the other way, because of ideological sympathy, political donations, or the revolving door. Additionally, flaws in contract mechanisms or oversight invite, or at least overlook, abuse, the DoD is audited imperfectly, and some military and intelligence programs put astounding amounts of money in play with poor accountability.

42. J. Paul Dunne and Nan Tian, "Military Expenditure and Economic Growth," *The Economics of Peace and Security Journal* 8, no. 1 (2013): 5–11, quote at p. 8.

43. J. Paul Dunne and Elisabeth Sköns, "The Military Industrial Complex," in Tan, ed., *The Global Arms Trade*, 281–92.

44. J. Paul Dunne and Elizabeth Sköns, "The Changing Military-Industrial Complex," *Working Papers* 1104, Department of Accounting, Economics and Finance, Bristol Business School, University of the West of England, Bristol (2011), 7.

Selected Bibliography

Adams, Gordon. *The Politics of Defense Contracting: The Iron Triangle.*
New Brunswick, NJ: Transaction Books, [1981] 1982.

Alic, John A. *Trillions for Military Technology: How the Pentagon Innovates and Why It Costs So Much.* New York: Palgrave Macmillan, 2007.

Alic, John A., Lewis M. Branscomb, Harvey Brooks, Ashton B. Carter, and Gerald L. Epstein. *Beyond Spinoff: Military and Commercial Technologies in a Changing World.* Boston, MA: Harvard Business School Press, 1992.

Bacevich, Andrew W., ed. *The Long War: A New History of U.S. National Security Policy since World War II.* New York: Columbia University Press, 2007.

Bardes, Barbara, and Robert W. Oldendick. *Public Opinion: Measuring the American Mind,* 5th ed. Lanham, MD: Rowman & Littlefield, 2017.

Beard, Edmund. *Developing the ICBM: A Study in Bureaucratic Politics.* New York: Columbia University Press, 1976.

Benviste, Meron, ed. *Abu Ghraib: The Politics of Torture.* London: Atlantic Books, 2004.

Bingham, Richard D. *Industrial Policy American Style: From Hamilton to HDTV.* Armonk, NY: M. E. Sharpe, 1998.

Bilker, John, John F. Schank, Giles K. Smith, Fred Timson, James Chiesa, Marc Goldberg, Michael G. Mattock, and Malcolm MacKinnon. *The U.S. Submarine Production Base: An Analysis of Cost, Schedule, and Risk for Selected Force Structures.* Santa Monica, CA: RAND, 1994.

Bitzinger, Richard A., ed. *The Modern Defense Industry.* Santa Barbara, CA: ABC-CLIO, 2009.

Block, Fred, and Matthew R. Keller. "Where Do Innovations Come From?

Transformations in the US Economy, 1970–2006." *Socio-Economic Review* 7 (2009): 459–83.

Boyle, Michael J. *The Drone Age: How Drone Technology Will Change War and Peace.* New York: Oxford University Press, 2020.

Bridger, Sarah. *Scientists at War: The Ethics of Cold War Weapons Research.* Cambridge, MA: Harvard University Press, 2015.

Brose, Christian. "The New Revolution in Military Affairs." *Foreign Affairs* (May/June 2019): 122–34.

Chatterjee, Pratap. *Halliburton's Army: How a Well-Connected Texas Oil Company Revolutionized the Way America Makes War.* New York: Nation Books, 2009.

Coburn, Noah. *Under Contract: The Invisible Workers of America's Global War.* Stanford, CA: Stanford University Press, 2018.

Congressional Budget Office. *Funding for Overseas Contingency Operations and Its Impact on Defense Spending.* CBO Report 54219 (Oct. 2018), 1, https://www.cbo.gov/system/files/2018-10/54219-oco _spending.pdf.

Cordesman, Anthony. "U.S. Military Spending: The Cost of Wars." Center for Strategic and International Studies (July 10, 2017), https://www .csis.org/burke/reports.

Cox, Ronald W. "The Military-Industrial Complex and US Military Spending After 9/11." *Class, Race and Corporate Power* 2, no. 2 (2014), https://digitalcommons.fiu.edu/classracecorporatepower/vol2 /iss2/5/.

DeGrasse, Robert W., Jr. *Military Expansion, Economic Decline: The Impact of Military Spending on U.S. Economic Performance.* Armonk, NY: M. E. Sharpe, 1983.

Deutch, John. "Consolidation of the U.S. Defense Industrial Base." *Acquisitions Review Quarterly* 8, no. 3 (Fall 2001): 137–50.

Duncan, Thomas K., and Christopher J. Coyne. "The Origins of the Permanent War Economy." *The Independent Review* 18, no. 2 (Fall 2013): 219–40.

Dunne, J. Paul, and Elizabeth Sköns. "The Changing Military-Industrial Complex." *Working Papers* 1104. Department of Accounting, Econom-

ics and Finance, Bristol Business School, University of the West of
England, 2011.

———. "The Military-Industrial Complex." In *The Global Arms Trade:
A Handbook*, ed. Andrew T. H. Tan, 281–92. New York: Routledge,
2010.

Dunne, J. Paul, and Nan Tian. "Military Expenditure and Economic Growth."
The Economics of Peace and Security Journal 8, no. 1 (2013): 5–11.

Dunne, J. Paul, and Mehmet Uye. "Defense Spending and Development."
In *The Global Arms Trade: A Handbook*, ed. Andrew T. H. Tan,
293–305. New York: Routledge, 2010.

Eloranta, Jari. "Military Spending Patterns in History." In *EH.Net
Encyclopedia*, ed. Robert Whaples (Sept. 16, 2005), https://eh
.net/?s=Jari+Eloranta+Military+spending+patterns+in+history.

Fallows, James. "The Military-Industrial Complex." *Foreign Policy* 133
(Nov.–Dec. 2002): 46–48.

Farrell, Theo. "Weapons without a Cause: Buying Stealth Bombers the
American Way." *Contemporary Security Policy* 14, no. 2 (1993):
115–50.

———. *Weapons without a Cause: The Politics of Weapons Acquisition in
the United States*. New York: St. Martin's Press, 1997.

Feaver, Peter D. *Armed Servants: Agency, Oversight, and Civil-Military
Relations*. Cambridge, MA: Harvard University Press, 2003.

Ferris, John. "Netcentric Warfare, C4ISR and Information Operations:
Towards a Revolution in Military Intelligence?" *Intelligence and
National Security* 19, no. 2 (Summer 2004): 199–225.

Fine, Ben. "The Military-Industrial Complex: An Analytical Assessment."
Cyprus Journal of Economics 6, no. 1 (June 1993): 26–51.

Foerstel, Herbert N. *Secret Science: Federal Control of American Science
and Technology*. Westport, CT: Praeger, 1993.

Fordham, Benjamin O. "Paying for Global Power: Assessing the Costs and
Benefits of Postwar U.S. Military Spending." In *The Long War: A New
History of US National Security Policy since World War II*, ed. Andrew
Bacevich, 371–404. New York: Columbia University Press, 2007.

Fox, J. Ronald, David G. Allen, Thomas C. Lassman, Walton S. Moody, and Philip L. Shiman. *Defense Acquisition Reform, 1960–2009: An Elusive Goal*. Washington, DC: Center of Military History, United States Army, 2001.

Friedberg, Aaron L. *In the Shadow of the Garrison State: America's Anti-Statism and its Cold War Grand Strategy*. Princeton, NJ: Princeton University Press, 2000.

———. "Why Didn't the United States Become a Garrison State?" *International Security* 16, no. 4 (Spring 1992): 109–42.

Gaddis, John Lewis. "A Grand Strategy of Transformation." *Foreign Policy* 133 (Nov.–Dec. 2002): 50–57.

Gansler, Jacques S. *The Defense Industry*. Cambridge, MA: MIT Press, 1980.

———. *Defense Conversion: Transforming the Arsenal of Democracy*. Cambridge, MA: MIT Press, 1995.

———. *Democracy's Arsenal: Creating a Twenty-First Century Defense Industry*. Cambridge. MA: MIT Press, 2011.

Gansler, Jacques S., William Lucyshyn, and William Varettoni. *Acquisition of Mine-Resistant, Ambush-Protected (MRAP) Vehicles: A Case Study*. School of Public Policy, University of Maryland: NPS Acquisition Research Symposium, May 12, 2010.

Gertler, Jeremiah. *F-35 Joint Strike Fighter (JSF) Program*. CRS Report RL30563 (Apr. 23, 2018).

Gholz, Eugene. "The Curtiss-Wright Corporation and Cold War-Era Defense Procurement: A Challenge to Military-Industrial Complex Theory." *Journal of Cold War Studies* 2, no. 1 (Winter 2000): 35–75.

———. "Eisenhower versus the Spin-Off Story: Did the Rise of the Military–Industrial Complex Hurt or Help America's Commercial Aircraft Industry?" *Enterprise & Society* 12, no. 1 (Mar. 2011): 46–95.

Gibney, Frank. "The Missile Mess." *Harper's Magazine* 220 (Jan. 1, 1960): 38–45.

Hansen, Morton. "Intelligence Contracting: On the Motivations, Interests, and Capabilities of Core Personnel Contractors in the US Intelligence Community." *Intelligence and National Security* 29, no. 1 (2014): 58–81.

Harris, Shane. *@War: The Rise of the Military-Internet Complex*. Boston: Houghton Mifflin, 2014.

Hart, Gary, and William S. Lind. *America Can Win: The Case for Military Reform*. Bethesda, MD: Adler & Adler, 1986.

Hartung, William D. *And Weapons for All*. New York: HarperCollins, 1994.

———. *Prophets of War: Lockheed Martin and the Making of the Military-Industrial Complex*. New York: Nation Books, 2011.

Hasik, James. *Arms and Innovation: Entrepreneurship and Alliances in the Twenty-First Century Arms Industry*. Chicago: University of Chicago Press, 2008.

———. *Securing the MRAP: Lessons Learned in Marketing and Military Procurement*. College Station: Texas A&M University Press, 2021.

———. "Beyond the Third Offset: Matching Plans for Innovation to a Theory of Victory," *Joint Forces Quarterly* 91 (4th Quarter, 2018): 14–21.

Higgs, Robert. "The Cold War Economy: Opportunity Costs, Ideology, and the Politics of Crisis," *Explorations in Economic History* 31 (1994): 283–312.

———. "Profits of U.S. Defense Contractors." The Independent Institute, Mar. 13, 1992, https://www.independent.org/publications/article.asp?id=129.

———. "The Trillion-Dollar Defense Budget Is Already Here." The Independent Institute, Mar. 15, 2007, www.independent.org/news/article.asp?id=1941.

———. "U.S. Military Spending in the Cold War Era: Opportunity Costs, Foreign Crises, and Domestic Constraints." Cato Institute Policy Analysis No. 114, Nov. 30, 1988, https://www.object.cat.org/pubs/pas/pa114.pdf.

Hiltzik, Michael. *Big Science: Ernest Lawrence and the Invention that Launched the Military-Industrial Complex*. New York: Simon & Schuster, 2015.

Hogan, Michael J. *A Cross of Iron: Harry S. Truman and the National Security State, 1945–1954*. New York: Cambridge University Press, 1998.

Hounshell, David. "The Evolution of Industrial Research in the United States." In *Engines of Innovation: U.S. Industrial Research at the End of an Era*, ed. R. Rosenbloom and W. J. Spencer, 13–85. Cambridge, MA: Harvard Business School Press, 1996.

Hubinger, Scott. "Can the F-35 Lightning II Joint Strike Fighter Avoid the Fate of the F-22 Raptor?" *Joint Force Quarterly* 94 (3rd Quarter 2019), 44–52.

Hughes, Michael P. "Lockheed Martin and the Controversial F-35." *Journal of Business Case Studies* 11, no. 1 (1st Quarter 2015): 1–14.

Isaacson, Walter. "The Winds of Reform." *Time*, Mar. 7, 1983, http://content.time.com/time/magazine/article/0,9171,953733,00.html.

Janiewski, Delores E. "Eisenhower's Paradoxical Relationship with the "Military-Industrial Complex." *Presidential Studies Quarterly* 4, no. 41 (Dec. 2011): 667–93.

Kennedy, Gavin. *Defense Economics*. New York: St. Martin's Press, 1983.

Koistinen, Paul A. C. *State of War: The Political Economy of American Warfare, 1945–2011*. Lawrence: University of Kansas, 2012.

Kopp, Carlo. "Lockheed Martin's F-35 Joint Strike Fighter." *Australian Aviation* (Apr.–May 2002): 24–32, http://ausairpower.net/jsf-analysis-2002.html.

Kosiak, Steven M. *Is the U.S. Military Getting Smaller and Older? And How Much Should We Care?* Washington, DC: Center for a New American Security, 2017.

Kreps, Sarah. *Drones: What Everyone Needs to Know*. New York: Oxford University Press, 2016.

LaFeber, Walter. "The Rise and Fall of Colin Powell and the Powell Doctrine." *Political Science Quarterly* 124, no. 1 (Spring 2009): 71–93.

Ledbetter, James. *Unwarranted Influence: Dwight Eisenhower and the Military-Industrial Complex*. New Haven, CT: Yale University Press, 2011.

Light, Paul C. "Fact Sheet on the True Size of Government." Accessed Aug. 27, 2019. www.wagner.nyu.edu/files/faculty/publications/lightFactTrueSize.pdf.

———. *The Government-Industrial Complex: The True Size of the Federal Government, 1984–2018*. New York: Oxford University Press, 2019.

———. "The True Size of Government: Tracking Washington's Blended Workforce, 1984–2015." The Volker Alliance, Issue Papers, Oct. 5, 2017, https://www.volckeralliance.org/publications/true-size-government.

Lindsay, James M. "Parochialism, Policy, and Constituency Constraints: Congressional Voting on Strategic Weapons Systems." *American Journal of Political Science* 34, no. 4 (Nov. 1990): 936–60.

Lonnquest, John Clayton. "The Face of Atlas: General Bernard Schriever and the Development of the Atlas Intercontinental Ballistic Missile, 1953–1960." PhD diss., Duke University, 1996.

Lundquist, Jerrold T. "Shrinking Fast and Smart." *Harvard Business Review* (Nov.–Dec. 1992): 74–85.

Lynn, William J., III. "The End of the Military-Industrial Complex." *Foreign Affairs* 93, no. 6 (Nov.–Dec. 2014): 104–10.

Mahnken, Thomas G. *Technology and the American Way of War*. New York: Columbia University Press, 2008.

Mann, James. *The Rise of the Vulcans: The History of Bush's War Cabinet*. New York: Viking, 2004.

Markusen, Ann. "Dismantling the Cold War Economy." *World Policy Journal* 9, no. 3 (Summer 1992): 389–99.

Markusen, Ann, and Joel Yudken. *Dismantling the Cold War Economy*. New York: Basic Books, 1992.

Martin, Donna. "Defense Procurement Information Papers: Campaign '84." Project on Military Procurement. Washington, DC: Fund for Constitutional Government, Aug. 1984.

Mathews, Jessica T. "America's Indefensible Defense Budget." *The New York Review of Books* (July 18, 2019), www.nybooks.com/articles /2019/07/18/americas-indefensible-defense-budget/.

Mayer, Kenneth R. *The Political Economy of Defense Contracting*. New Haven, CT: Yale University Press, 1991.

McCartney, James, and Molly Sinclair McCartney. *America's War Machine: Vested Interests, Endless Conflicts*. New York: St. Martin's, 2015.

McDougall, Walter A. *The Tragedy of U.S. Foreign Policy: How America's Civil Religion Betrayed the National Interest.* New Haven, CT: Yale University Press, 2016.

McGarry, Brenda W., and Emily M. Morgenstern. *Overseas Contingency Operations Funding: Background and Status.* CRS Report 44519 (Sept. 6, 2019).

McInnes, Kathleen. *Goldwater-Nichols at 30: Defense Reform and Issues for Congress.* CRS Report R44474 (June 2, 2016).

Mearsheimer, John J. "The Military Reform Movement: A Critical Assessment." *Orbis* 27, no. 2 (Summer 1983): 285–300.

Miles, Anne Daugherty. *Intelligence Community Programs, Management, and Enduring Issues.* CRS Report R44681 (Nov. 8, 2016).

———. *Intelligence Spending: In Brief.* CRS Report 44381 (Feb. 16, 2016).

Mintz, Alex. "The Military-Industrial Complex: American Concepts and Israeli Realities." *Journal of Conflict Resolution* 29, no. 4 (Dec. 1985), 623–39.

———, ed. *The Political Economy of Military Spending in the United States.* London: Routledge, 1992.

Molzahn, Wendy. "The CIA's In-Q-Tel Model: Its Applicability." *Acquisition Review Quarterly* 10, no. 1 (Winter 2003): 46–61.

Mowery, David C., and Nathan Rosenberg. *Technology and the Pursuit of Economic Growth.* Cambridge: Cambridge University Press, 1994.

National Science Foundation, National Center for Science and Engineering Statistics. *Federal R&D Funding, by Budget Function: Fiscal Years 2016–18, Detailed Statistical Tables,* NSF 18–308 (June 2018).

———. *National Patterns of R&D Resources: 2016–17 Data Update,* NSF19-309 (Feb. 27, 2019).

Nelson, Anna Kasten. "The Evolution of the National Security State" in *The Long War: A New History of U.S. National Security Policy since World War II,* ed. Andrew Bacevich, 265–301. New York: Columbia University Press, 2007.

Niemi, Christopher J. "The F-22 Acquisition Program: Consequences for the U.S. Air Force's Fighter Fleet." *Air & Space Power Journal* 26, no. 6 (Nov.–Dec. 2012): 53–82.

Noble, David. *America by Design: Science, Technology, and the Rise of Corporate Capitalism*. New York: Knopf, 1977.

O'Hanlon, Michael E. "A Retrospective on the So-called Revolution in Military Affairs, 2000–2020." Brookings Report, Sept. 2018, https://www.brookings.edu/research/a-retrospective-on-the-so-called-revolution-in-military-affairs-2000-2020/.

———. *The Senkaku Paradox: Risking Great Power War over Small Stakes*. Washington, DC: Brookings Institution Press, 2019.

———. *Technological Change and the Future of Warfare*. Washington, DC: Brookings Institution Press, 2000.

———. "U.S. Defense Strategy and the Defense Budget." Brookings Institution, Budgeting for National Priorities Project, Nov. 18, 2015, www.brookings.edu/research/u-s-defense-strategy-and-the-defense-budget/.

O'Rourke, Ronald. *Navy Virginia (SSN-774) Class Attack Submarine Procurement: Background and Issues for Congress*. CRS Report RL32418 (June 4, 2019).

———. *V-22 Osprey Tilt-Rotor Aircraft: Background and Issues for Congress*. CRS Report 31384 (June 10, 2009).

Peters, Heidi M. *Department of Defense Use of Other Transaction Authority: Background, Analysis, and Issues for Congress*. CRS Report R45521 (Feb. 22, 2019).

Peters, Heidi M., Moshe Schwartz, and Lawrence Kapp. *Department of Defense Contractor and Troop Levels in Iraq and Afghanistan: 2007–2017*. CRS Report 45521 (Apr. 28, 2017).

Pierre, Andrew J., ed. *Cascade of Arms: Managing Conventional Weapons Proliferation*. Washington, DC: Brookings Institution Press, 1997.

Pike, John. "Mine Resistant Ambush Protected (MRAP) Vehicle Program." Accessed Aug. 5, 2019. www.globalsecurity.org/military/systems/ground/MRAP.htm.

Priest, Dana, and William M. Arkin. *Top Secret America: The Rise of the New American Security State*. New York: Little, Brown, 2011.

Rid, Thomas. *Cyber War Will Not Take Place*. New York: Oxford University Press, 2013.

Roland, Alex. "Fermis as the Measure of War: Neutrons, Photons,

Electrons and the Sources of Military Power." In *Tooling for War: Military Transformation in the Industrial Age*, ed. Stephen D. Chiabotti, 173–88. Chicago: Imprint Publications, 1996.

———. "The Grim Paraphernalia: Eisenhower and the Garrison State." In *Forging the Shield: Eisenhower and National Security in the 21st Century*, ed. Dennis Showalter, 13–22. Carson City, NV: Imprint Publications, 2005.

———. "Is Military Technology Deterministic?" *Vulcan* 7 (2019): 19–33.

———. "Keep the Bomb," *Technology Review* (Aug.–Sept. 1995): 67–69.

———. *The Military-Industrial Complex*. Washington, DC: American Historical Association, 2001.

———. "The Shuttle: Triumph or Turkey?" *Discover* 6 (Nov. 1985): 29–49.

Roland, Alex, with Philip Shiman. *Strategic Computing: DARPA and the Quest for Machine Intelligence, 1983–1993*. Cambridge, MA: MIT Press, 2002.

———. "Was the Nuclear Arms Race Deterministic?" *Technology and Culture* 51 (April 2010): 444–61.

Steven Rosen, ed. *Testing the Theory of the Military-Industrial Complex*. Lexington, MA: D. C. Heath, 1973.

Rumsfeld, Donald. "Transforming the Military." *Foreign Affairs* 81, no. 3 (May–June 2002), 20–32.

Rundquist, Barry S., and Thomas M. Casey. *Congress and Defense Spending: The Distributive Politics of Military Procurement*. Norman: University of Oklahoma Press, 2002.

Ruttan, Vernon W. *Is War Necessary for Economic Growth? Military Procurement and Technology Development*. New York: Oxford University Press, 2006.

Sampson, Anthony. *The Arms Bazaar: From Lebanon to Lockheed*. New York: Viking, 1977.

Samuels, Richard J. *Rich Nation, Strong Army: National Security and the Technological Transformation of Japan*. Ithaca, NY: Cornell University Press, 1994.

Sandler, Todd, and Keith Hartley, *The Economics of Defense*. Cambridge: Cambridge University Press, 1995.

Sapolsky, Harvey M. *Polaris System Development: Bureaucratic and Programmatic Success in Government.* Cambridge, MA: Harvard University Press, 1972.

Sargent, John F., Jr., *U.S. Research and Development Funding and Performance Fact Sheet.* CRS Report R44307 (Sept. 19, 2019).

Sargent, John F., Moshe Schwartz, and Marcey E. Gallo. *The Global Research and Development Landscape and Implications for the Department of Defense.* CRS Report 45403 (Nov. 8, 2018).

Schank, John F., Cesse Ip, Frank W. Lacroix, Robert Murphy, Mark V. Arena, Kristy N. Kamarck, and Gordon T. Lee. *Learning from the U.S. Navy's* Ohio, Seawolf, *and* Virginia *Submarine Programs,* Vol II of *Learning from Experience.* Santa Monica, CA: RAND Corporation, 2011.

Schóber, T., and P. Puliš. "F-35—Win or Loss for the USA and Their Partners?" *Advances in Military Technology* 10, no. 2 (Dec. 2015): 82–94.

Schwartz, Stephen I., and Deepti Choosey. *Nuclear Security Spending: Assessing Costs, Examining Priorities.* Washington, DC: Carnegie Endowment for International Peace, 2009.

Shachtman, Noah. "Exclusive: Google, CIA Invest in 'Future' of Web Monitoring." Wired.com, July 28, 2010, www.wired.com/2010/07/exclusive-google-cia/.

Sherry, Michael S. *In the Shadow of War: The United States since the 1930s.* New Haven, CT: Yale University Press, 1995.

Shorrock, Tim. *Spies for Hire: The Secret World of Intelligence Outsourcing.* New York: Simon & Schuster, 2008.

Simon, Nora, and Alain Minc. *The Computerization of Society: A Report to the President of France.* Cambridge, MA: MIT Press, 1980.

Singer, P. W. *Corporate Warriors: The Rise of the Privatized Military Industry.* Ithaca, NY: Cornell University Press, 2003.

———. "The Dark Truth about Blackwater." *Brookings,* Oct. 2, 2007, www.brookings.edu/articles/the-dark-truth-about-blackwater/.

———. *Wired for War: The Robotics Revolution and Conflict in the 21st Century.* New York: Penguin, 2009.

Sköns, Elisabeth. "The U.S. Defense Industry after the Cold War." In *The

Global Arms Trade: A Handbook, ed. Andrew T. H. Tan, 235–49. New York: Routledge, 2010.

Smith, Ron. *Military Economics: The Interaction of Power and Money*. New York: Palgrave Macmillan, 2009.

Sowell, P. Kathie. "The C4ISR Architecture Framework: History, Status, and Plans for Evolution." The MITRE Corporation, *Technical Papers*, Aug. 2000, https://www.mitre.org/publications/technical-papers/the -c4isr-architecture-framework-history-status-and-plans-for-evolution.

Stockholm International Peace Research Institute (SIPRI). "SIPRI Arms Transfer Database." Accessed Jan. 29, 2020, www.sipri.org/databases /armstransfers/.

Tan, Andrew T. H., ed. *The Global Arms Trade: A Handbook*. New York: Routledge, 2010.

Thompson, Loren. "How the Pentagon's Top Ten Contractors Dealt with the Last Downturn." *Forbes*, Feb. 13, 2013, https://www.forbes.com /sites/lorenthompson/2013/02/13/how-the-pentagons-top-ten -contractors-dealt-with-the-last-downturn/#27f0822859a2.

Thompson, Mark. "The Most Expensive Weapon Ever Built." *TIME Magazine* 181, no. 7, Feb. 25, 2013, http://content.time.com/time /magazine/article/0,9171,2136312,00.html.

Thorpe, Rebecca U. *The American Warfare State: The Domestic Politics of Military Spending*. Chicago: University of Chicago Press, 2014.

Turse, Nick. *The Complex: How the Military Invades Our Everyday Lives*. London: Faber & Faber, 2009.

Tyler, Patrick E. "U.S. Strategy Plan Calls for Insuring No Rivals Develop." *New York Times*, Mar. 8, 1992, https://www.nytimes.com/1992/03/08 /world/us-strategy-plan-calls-for-insuring-no-rivals-develop.html.

United States, White House, Office of Management and Budget [OMB]. *Historical Tables*. Accessed Oct. 21, 2019. www.whitehouse.gov/omb /historical-tables.

United States Government Accountability Office. "High Risk Series." Report to Congressional Committees. GAO-19-157SP (March 2019). Washington, DC.

———. *Weapon Systems Annual Assessment: Limited Use of Knowledge-*

Based Practices Continues to Undercut DOD's Investments. GAO-19-336SP (May 2019). Washington, DC.

US Congress, Office of Technology Assessment. *After the Cold War: Living with Lower Defense Spending.* Washington, DC: Congress of the United States, Office of Technology Assessment, February 1992.

———. *Redesigning Defense: Planning the Transition to the Future U.S. Defense Industrial Base.* OTA-ISC-500 (July 1991). Washington, DC: Government Printing Office.

———. *Weapon Systems Annual Assessment: Limited Use of Knowledge-Based Practices Continues to Undercut DOD's Investments.* GAO-19-336SP (May 2019). Washington, DC: Government Printing Office.

US Congressional Budget Office. *Funding for Overseas Contingency Operations and Its Impact on Defense Spending.* CBO Report 54219 (Oct. 2018), www.cbo.gov/system/files/2018-10/54219-oco_spending.pdf.

Vucetic, Srdjan, and Kim Richard Dossal. "The International Politics of the F-35 Joint Strike Fighter." *International Journal* (Winter 2012–13): 3–12.

Wang, Jessica. *American Science in an Age of Anxiety: Scientists, Anticommunism, and the Cold War.* Chapel Hill: University of North Carolina Press, 1999.

Weber, Rachel. *Swords into Dow Shares: Governing the Decline of the Military-Industrial Complex.* Boulder, CO: Westview Press, 2001.

Weinberger, Sharon. *The Imagineers of War: The Untold Story of DARPA, the Pentagon Agency that Changed the World.* New York: Knopf, 2017.

Weiss, Linda. *America Inc.? Innovation and Enterprise in the National Security State.* Ithaca, NY: Cornell University Press, 2014.

———. "Crossing the Divide: From the Military-Industrial to the Development-Procurement Complex." Paper presented at the Berkeley Workshop on the "Hidden US Development State," San Francisco, CA, June20–21, 2008, https://www.researchgate.net/publication/2287 11954_Crossing_the_Divide_From_the_Military-Industrial_to_the _Development-Procurement_Complex.

Winternitz, Luke. *Introduction to GPS and Other Global Navigation*

Satellite Systems. NASA/Goddard Space Flight Center Code 596, presented at 42nd Annual Time and Frequency Meteorology Seminar, June 8, 2017, www.ntrs.nasa.gov/archive/nasa/casi.ntrs.nasa.gov /20170004590.pdf.

Westwick, Peter J. *The National Labs: Science in an American System, 1947–1974.* Cambridge, MA: Harvard University Press, 2003.

Whittle, Richard. *Dream Machine: The Untold History of the Notorious V-22 Osprey.* New York: Simon & Schuster, 2010.

Wilson, Mark. "The Military-Industrial Complex." In *At War: The Military and American Culture in the Twentieth Century and Beyond,* ed. David Kieran and Edwin A Martini, 67–86. New Brunswick, NJ: Rutgers University Press, 2018.

Wright, Lawrence. "The Spymaster." *The New Yorker* 83, no. 44, Jan. 21, 2008, https://www.newyorker.com/magazine/2008/01/21/the-spy master.

Yergen, Daniel. *Shattered Peace: The Origins of the Cold War and the National Security State.* New York: Houghton Mifflin, 1977.

Zaffran, Raphaël, and Nicolas Ewes. "Beyond the Point of No Return: Allied Defense Procurement, the 'China Threat,' and the Case of the F-35 Joint Strike Fighter." *Asian Journal of Public Affairs* 8, no. 1 (2015): 64–88.

Zegart, Amy B. "An Empirical Analysis of Failed Intelligence Reforms before September 11." *Political Science Quarterly* 121, no. 1 (Spring 2006): 33–60.

Index

A page number in italics indicates an image or a chart. The letters "gl" indicate a glossary entry.

Bush, George W.: and defense reform, 130–31; and the MIC, 138; and RMA, 138; War on Terror, 133–36

Bush, Vannevar, 50, 51, 120, 122, 192, 222n11

buying-in, 33, 34, 64, 128, 151, 195–96, 213gl

C³I (Command, control, communication, and information), 80, 213gl, 233n13

C⁴ISR (Command, Control, Communication, Computers, Intelligence, Surveillance, and Reconnaissance), 172, 212, 213gl, 233n13, 253n19

C-5A transport aircraft, 57–58, 129, 173, 213gl, 230n16

C-135 aircraft, 55

CACI (aka Consolidated Analysis Center, Inc.), 148–49

"capability greed," 38, 151, 213gl, 227n56

capitalist marketplace, 4, 6, 54, 124, 225n40

Carter, Jimmy, 31, 34

casualties, military: Iraq war, 140–41, 245n13; minimizing, 18, 107, 109, 142–43, 159, 160, 190; and networked infrastructure, 194

Central Intelligence Agency (CIA). See CIA

Cheney, Dick, 56, 131, 139

CIA: and Abu Ghraib, 158–59, 251n68; in Afghanistan, 137; birth of, 28; and the IIC, 135, 150; in Iran, 94; and Iraq invasion, 162; and national military establishment, 5–6; outsourcing, 243n72; politicization of, 61, 62; and "venture innovation," 125, 146; and web monitoring, 243n68

civil/military relations: blurring of boundaries, 147–48, 155, 161, 174, 183, 188, 190, 202; civilian control of military, 40, 41, 44, 45, 155, 174, 227n1, 261n29; and intelligence industry, 174; and MIC autonomy, 34–35; in new world order, 205; public/private divide and civil society, 204–5

classification, security, 37, 69–70, 71, 204

classified materials, 29, 45–46, 70, 145, 160

Clinton administration, 95, 116–17, 129, 130

Cold War: legacy weapon systems, 108, 110–11, 111, 132, 166–67, 185, 212; paradigm, 104, 107–8, 132

collusion, 9–10, 37–38, 55, 57–58, 127

commercial-off-the-shelf (COTS), 48, 125–26, 144, 148, 150, 161, 205, 212

commonality, 41–42, 177, 178, 189

communications: and computer technology, 80; equipment, 84, 140, 142, 144; innovations in, 254n21; interceptions, 160, 200; and National Security Agency (NSA), 62; and net-centric warfare, 137; and the Patriot Act, 135; and space, 28; US technological strengths in, 103, 107

communism: and command economy, 39, 54, 78; movement (world communism), 4, 24; threat of expansion, 24, 35, 41, 45, 73

computer technology: C³I and the Cold War, 233n12; and DARPA, 47–48; as "defining technology," post-Cold War, 104–6; and dual-use applications, 82, 209; government support for, 80, 243n67, 243n69; loss of human agency and, 72, 78–79, 193; and technology transfer, 84; "thick fabric," 54–55, 62

conflict of interest, 53, 54, 55, 198, 204

VTOL/STOL aircraft, 112. *See also* Osprey
  aircraft (V-22 Osprey)
Vulcans, 131, 137, 139, 217gl, 245n7

War on Terror, Global: Afghanistan
  cam-paign, 136–37; arsenal, 166–67;
  budget, 145, 163, 164, 169; disenchantment with, 204; Homeland Security, Department of (DHS), 135–36,
  148–51, 190, 191; Iraq invasion, 162;
  and Obama administration, 167–69;
  and OODA loop, 238n17; "Overseas
  Contingency Operations," 163, 168,
  169; Patriot Act (2001), 135
Warsaw Pact, 83–84, 86
weapons program, 65–66, 196, 234n3
weapons sales. *See* arms trade; arms
  transfers
weapon systems: acquisitions, 196–97,
  200–201; and advanced technology,
204; atomic, 16, 25, 64, 233n10,
246n20; "capability greed," 38, 151,
213gl, 227n56; Crusader artillery
system, 111; gold-plating, 114, 128,
151, 172, 197, 214gl; layered defense
system, 77; legacy (Cold War), 108,
110–11, 111, 132, 166–67, 185, 212;
nuclear, 43, 60, 71–73, 85, 96, 98;
smart, 37, 80; strategic, 76–77. *See also*
ballistic missiles
Weinberger-Powell Doctrine, 188, 252n6
Weiss, Linda, 105–6, 170, 204, 205,
  209–10
Wilson, Charles E., 22, 35
Wolfowitz, Paul, 97, 137, 139
Wooldridge, Dean, 52
world GDP, 99, 211
World War I, 15, 16, 17
World War II, 4, 5, 6, 16, 17
World War III, 16, 75